Springer Proceedings in Mathematics & Statistics

Volume 511

This book series features volumes composed of selected contributions from workshops and conferences in all areas of current research in mathematics and statistics, including data science, operations research and optimization. In addition to an overall evaluation of the interest, scientific quality, and timeliness of each proposal at the hands of the publisher, individual contributions are all refereed to the high quality standards of leading journals in the field. Thus, this series provides the research community with well-edited, authoritative reports on developments in the most exciting areas of mathematical and statistical research today.

Alejandra Avalos-Pacheco · Fan Bu ·
Beatrice Franzolini · Beniamino Hadj-Amar
Editors

New Trends in Bayesian Statistics

BAYSM 2023, Online Meeting, November 13–17, Selected Contributions

Springer

Editors
Alejandra Avalos-Pacheco
Johannes Kepler University Linz
Linz, Austria

Fan Bu
University of Michigan
Ann Arbor, MI, USA

Beatrice Franzolini
Università Bocconi
Milan, Italy

Beniamino Hadj-Amar
Rice University
Houston, TX, USA

ISSN 2194-1009 ISSN 2194-1017 (electronic)
Springer Proceedings in Mathematics & Statistics
ISBN 978-3-031-99008-3 ISBN 978-3-031-99009-0 (eBook)
https://doi.org/10.1007/978-3-031-99009-0

Mathematics Subject Classification: 62H12, 62F15, 62G05

This Springer imprint is published by the registered company Springer Nature Switzerland AG
The registered company address is: Gewerbestrasse 11, 6330 Cham, Switzerland

Preface

In today's data-driven landscape, the challenges of modeling complex, high-dimensional, and dynamic phenomena call for statistical methods that not only deliver accurate predictions but also offer transparent uncertainty quantification and lower-dimensional interpretable representations of complex phenomena. In an era defined by the growth of available data and its inherent complexity, the need for statistical methods that can unravel high-dimensional, structured, and dynamic phenomena has never been more urgent. Bayesian statistics, with its formal mathematical framework for incorporating prior information and quantifying uncertainty, has emerged as a cornerstone of modern data analysis.

New Trends in Bayesian Statistics is a timely volume that not only includes some of the latest theoretical developments and computational innovations but also illustrates their transformative applications across diverse scientific domains. This book is designed to serve as both a reference and an inspiration, uniting methodological advances with practical challenges from biomedicine, economics, network science, and beyond. The present volume brings together a curated collection of research contributions that push the boundaries of Bayesian methods. At its heart, this book is a testament to the versatility of Bayesian inference in addressing some of the most pressing issues in contemporary data analysis. In particular, the contributions showcased here demonstrate how Bayesian approaches can seamlessly handle complex challenges such as high dimensionality, complex dependence structures, and dynamic behaviors—features that are increasingly prevalent in modern datasets. One of the central themes that run through the chapters is the interplay between rigorous theoretical underpinnings and innovative computational strategies.

The volume opens with the first chapter, "Segmenting Brain MALDI-MSI Data Under Separate Exchangeability". Mass spectrometry imaging (MSI) is revolutionizing biomedical research, enabling the in situ measurement of cancer biomarkers and critical biological traits. This chapter delves into the use of Bayesian models for nested, separately exchangeable data to analyze MSI data, specifically lipid abundance in mouse brain tissue. By employing a scalable coordinate ascent variational inference algorithm, the authors present a biclustering approach that uncovers meaningful patterns of molecule activation and image segmentation. This chapter bridges statistical innovation with biological discovery, addressing the challenges posed by high-dimensional data and computational efficiency.

Following this, second chapter, "A Bayesian Approach with Gaussian Priors to the Inverse Problem of Source Identification in Elliptic PDEs". Here, the focus shifts to inverse problems where the objective is to identify unknown source functions within elliptic partial differential equations (PDEs). The chapter illustrates how PDEs are central to modeling physical phenomena, yet identifying unknown source functions from noisy observations presents a challenging inverse problem. This chapter explores a Bayesian framework with Gaussian priors, leveraging conjugacy for posterior inference. Focusing on Gaussian series priors on the Dirichlet–Laplacian eigenbasis and Matérn process priors, the authors provide theoretical guarantees and implement posterior-based estimation for source identification. Through numerical simulations, the efficacy of these priors is demonstrated, offering a robust approach to uncertainty quantification in elliptic PDEs.

The third chapter, entitled "Phase I Dose Escalation Trials in Cancer Immunotherapy: Modifying the Bayesian Logistic Regression Model for Cytokine Release Syndrome", brings us directly into the realm of clinical applications. The context of cancer immunotherapy presents unique challenges in Phase I dose-escalation trials due to the potential onset of CRS. This chapter extends the Bayesian Logistic Regression Model (BLRM) to account for both CRS and non-CRS toxicities. The authors propose a methodology to identify optimal safe doses and dosing regimens that mitigate CRS risks while ensuring effective treatment. By generalizing the BLRM to model dose regimens and joint toxicity outcomes, this contribution advances dose-finding methodologies in the context of immunotherapy, offering a practical framework for clinical applications.

In the fourth chapter, "A Bayesian Multiple Ising Model", the focus moves towards the analysis of binary data and the inference of network structures. Graphical models are indispensable for uncovering conditional independence structures, yet binary data have been underexplored compared to Gaussian models. This chapter introduces a Bayesian framework for multiple Ising models, allowing for subgroup network analysis and variability due to external factors. The approach employs conjugate priors and Laplace approximations, with efficient model selection achieved through a Metropolis–Hastings algorithm. The proposed method balances network sparsity and edge selection while effectively identifying related subgroups. Applications demonstrate the method's versatility in learning subgroup structures in binary data, bridging gaps in the literature on multiple graphical models.

The exploration of dynamic phenomena takes center stage in the fifth chapter, "Bayesian Nonparametric Estimation of Time-Varying Macroeconomic Tail Risk". In today's volatile economic environment, the ability to assess tail risks—extreme events that can have devastating impacts—is of paramount importance. This chapter introduces a Bayesian nonparametric approach based on a time-dependent Dirichlet process mixture model, which captures the evolution of macroeconomic tail risks over time. The methodological innovations discussed here underscore the utility of Bayesian methods in providing nuanced assessments of risk in environments characterized by heavy-tailed and asymmetric data distributions. By extending the scope of Bayesian analysis to dynamic economic phenomena, this work complements earlier contributions by demonstrating that Bayesian methods are equally at home in both scientific and financial applications.

The sixth chapter, “A Metropolis–Hastings Algorithm for Sampling Coagulated Partitions”, addresses the growing concerns related to privacy-preserving data analysis. With the increasing need for anonymization techniques in modern data science, the authors develop an innovative algorithm for sampling from distributions of coagulated partitions. Despite the inherent sparsity of the target distribution, the proposed Metropolis–Hastings sampler achieves accurate approximations efficiently. This chapter is particularly significant as it highlights how Bayesian computational methods can be tailored to solve practical problems in privacy, thereby broadening the applicability of Bayesian models beyond conventional data analysis scenarios.

Finally, the seventh chapter, “Conditionally Partially Exchangeable Partitions for Dynamic Networks”, presents a novel framework for understanding evolving network structures. Dynamic networks, which are prevalent in fields ranging from social sciences to biology, pose unique challenges due to their time-varying nature. By extending the traditional stochastic block models through the incorporation of conditional partial exchangeability and hierarchical normalized completely random measures, this chapter provides a robust tool for detecting evolving communities within networks. This contribution not only reinforces the central role of Bayesian methods in network analysis but also ties together many of the themes discussed in earlier chapters, such as the need for adaptable models that can capture both static and dynamic dependencies.

Beyond the individual contributions, an underlying strength of this volume is its ability to connect disparate fields through common methodological themes. Whether addressing the nuances of biomedical imaging, the intricacies of inverse problems in physics, or the challenges of dynamic network modeling, each chapter emphasizes the importance of balancing theoretical rigor with computational feasibility. The volume gathers contributions presented at the Bayesian Young Statisticians Meeting (BAYSM) 2023, the official conference of j-ISBA, the junior section of the International Society for Bayesian Analysis, together with more invited papers from additional contributors. This prestigious event provides a platform for early-career researchers to showcase innovative work and engage in discussions with senior researchers that shape the future of Bayesian statistics. The feedback and collaborative spirit of BAYSM 2023 have left an indelible mark on the work presented in this volume, underscoring the dynamic and forward-thinking nature of the Bayesian research community. The inclusion of some additional contributions highlights the vibrancy and creativity of the next generation of Bayesian statisticians, offering a glimpse into cutting-edge methodologies and their diverse applications. As we delve deeper into this book, it becomes clear that the advancements in Bayesian statistics presented here are not merely incremental. Instead, they represent a significant evolution in our ability to model, predict, and interpret complex phenomena. The integration of novel computational techniques—such as advanced variational inference, tailored Metropolis—Hastings algorithms, and nonparametric methods—with robust theoretical frameworks is driving a new era of statistical innovation. These methods are increasingly critical in a world where the volume, variety, and velocity of data continue to grow at an unprecedented pace. Moreover, the interdisciplinary nature of the contributions serves as a reminder that modern statistical challenges do not respect traditional academic boundaries. The applications discussed range from clinical trial design in immunotherapy to risk assessment in macroeconomics, from the analysis of brain imaging data to the study of dynamic networks and privacy-preserving algorithms. This diversity not only highlights the wide applicability of Bayesian methods but also encourages cross-disciplinary collaborations that can lead to novel insights and breakthroughs.

In summary, *New Trends in Bayesian Statistics* offers a panoramic view of the current state-of-the-art in Bayesian theory, methodology, and applications. It encapsulates the spirit of innovation and collaboration that is driving the field forward, while also addressing the practical demands of analyzing complex, structured, and dynamic data. Whether you are a statistician seeking to deepen your understanding of Bayesian theory, a researcher looking for robust analytical tools, or a practitioner aiming to apply cutting-edge techniques to real-world problems, this volume provides a rich tapestry of ideas and approaches. The chapters that follow not only showcase individual methodological breakthroughs but also weave together a narrative of continuous improvement and interdisciplinary synergy. As you explore the following sections, we invite you to appreciate the profound impact of Bayesian statistics on modern data analysis and to envision the exciting future that lies ahead in this dynamic field.

Linz, Austria — Alejandra Avalos-Pacheco
Ann Arbor, USA — Fan Bu
Milan, Italy — Beatrice Franzolini
Houston, USA — Beniamino Hadj-Amar

Contents

Segmenting Brain MALDI-MSI Data Under Separate Exchangeability

Francesco Denti, Cecilia Balocchi, and Giulia Capitoli

Abstract Mass spectrometry imaging is emerging as a valuable tool for measuring in situ cancer biomarkers. In fact, it allows us to detect critical biological traits that would be overlooked with a simple visual morphological assessment of a sample. This technique measures the abundance of several specific molecules over multiple locations in a biological sample. For example, when analyzing a brain slide, it produces several profiles of molecule abundances, one for each analyzed pixel. The analysis of these complex data structures calls for developing tailored statistical methods. In this contribution, we propose employing models for nested, separately exchangeable data to estimate a biclustering solution, i.e., cluster locations characterized by similar abundance profiles, and simultaneously detect molecules with similar expressions within clusters of pixels. We analyze the brain image of a healthy mouse, comprising the mass spectra of approximately one hundred lipid molecules in approximately 1600 pixels. To address the large dimensionality of the dataset and the need for timely results, we apply a scalable coordinate ascent variational inference algorithm that dramatically scales the applicability of the model. We showcase how the estimated biclustering structure allows us to detect meaningful image segmentations and patterns of activated molecules.

Keywords Distributional clustering · Molecular abundance · Coordinate ascent variational inference

F. Denti (✉)
University of Padova, Padua, Italy
e-mail: francesco.denti@unipd.it

C. Balocchi
University of Edinburgh, Edinburgh, UK
e-mail: cecilia.balocchi@ed.ac.uk

G. Capitoli
University of Milan-Bicocca, Milan, Italy
e-mail: giulia.capitoli@unimib.it

A. Avalos-Pacheco et al. (eds.), *New Trends in Bayesian Statistics*,
Springer Proceedings in Mathematics & Statistics 511,
https://doi.org/10.1007/978-3-031-99009-0_1

1 Introduction

Over the last decade, *Matrix-Assisted Laser Desorption/Ionisation* (MALDI) Mass Spectrometry Imaging (MSI) has emerged as one of the key technologies for the in situ discovery of cancer biomarkers. This technology extracts the distribution of molecules such as lipids, peptides, and N-glycans in different locations of a biological sample [22]. Specifically, a slide of a biological sample is divided into a grid of pixels of 10 μm each. For each pixel, the instrument acquires a mass spectrum, which is a collection of molecular abundances (or intensities) measured at different values of the mass-to-charge index *m/z* (also known as signal) [6]. The three molecular classes could provide complementary information and characterize different histological regions; in this contribution, we focus on the *lipid* molecular abundances extracted from a healthy mouse brain tissue section, which is the first extracted in the sequential analysis [13].

We remark that our MALDI-MSI dataset had to be preprocessed using different steps to reduce the physiological instrumental and analytical variability. For more details about these essential preliminary steps, see [8]. The output of the preprocessing pipeline is a matrix where each row contains the molecular abundance for a specific *m/z* recorded across all the pixels; conversely, each column contains the abundance profile within a single pixel. In our study, the dimensions of the matrix we consider are approximately 80 by 1,600.

Our goal is to analyze these complex data to extract hidden features that can help us identify useful biomarkers and find brain areas characterized by similar patterns. As a first step, we can visualize some descriptive statistics to better understand the data. Figure 1 displays the mouse brain tissue section colored by the (normalized) values of the average, median, and variance of the distributions of lipid abundances computed for each pixel. These simple statistics already provide a glimpse of the multi-faceted complexity of the data. For example, inspecting the median map reveals a problematic area in the right part of the slide (in dark blue), a trait mainly overlooked by the other two statistics. After an ex-post biological assessment, the area of systematic

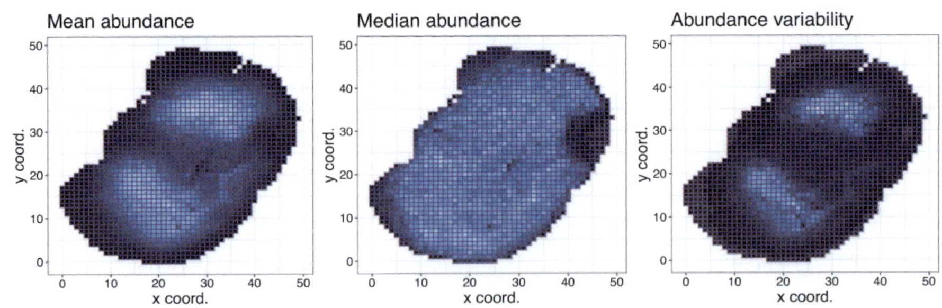

Fig. 1 Heatmaps representing, for each pixel, the mean, the median, and the variance of the distributions of the lipid abundances. For easier comparison, values are normalized, ranging from 0 (black) to 1 (light blue)

low abundance was revealed to be caused by a morphological defect of the sample, potentially being corrupted or too thin for a proper analysis. This example suggests that resorting to naive dimensionality reduction and summarizing the pixel-wise abundance distribution with a single statistic could lead to non-negligible information loss.

Hence, in what follows, we will model the molecular abundance profiles—one for each pixel—directly. We aim to reduce the complexity of the data through the detection of a small number of clusters, segmenting the image according to the distributional characteristics of the molecular signals. To achieve this goal, we employ a method to perform biclustering of the data, which allows us to simultaneously group the columns, i.e., the pixels, and, within each cluster of pixels, to group the rows, i.e., the different values of the mass-to-charge index *m/z*. Thus, the model we adopt induces a two-layer partition similar, for example, to the one discussed in [15] (see also [19] for other approaches to biclustering methods).

2 A Finite Separately Exchangeable Model for MALDI-MSI Data

Let $\boldsymbol{Y}$ indicate the MALDI-MSI matrix containing the measurements of interest, characterized by N rows (signals) and J columns (pixels). Our main aim is to perform image segmentation, i.e., clustering the J pixels driven by the characteristics of the molecular abundance profiles within each location. From a modeling perspective, we see our data as organized into distinct, albeit related, groups. This setting is typical of nested designs, an area of research that has thrived in recent years within the Bayesian literature [see, for example, 1, 2, 7, 12, 16, 21]. Most of these nested models were proposed under the hypothesis of partial exchangeability, where data are assumed to be exchangeable within the different subpopulations. However, Rebaudo et al. [20] have recently underlined that using a partial exchangeable framework to model matrix data disregards the information conveyed by the specificity of each row. Indeed, this feature is problematic for us, as each row of our matrix has the same meaning (i.e., it represents the same m/z value) across all the columns. In fact, in our framework using a partially exchangeable model would mean clustering the pixels that have similar abundance distributions, regardless of *which m/z* have high or low activation levels, and it would result in simply considering the proportion of *m/z* that have high or low activation levels. This might lead to an undesirable situation where, for example, two pixels are clustered together because they both have 5% of the signals that are highly activated, but maybe different sets of signals. Such a segmentation would lead to an important loss of information.

To solve this issue, the authors in [20] suggested adopting a separately exchangeable setting where the identity of each row is preserved across the groups. Thus, a more meaningful partition for our setting could be obtained by first introducing column-specific membership labels, denoted with $\boldsymbol{C} = (C_1, \ldots, C_J)$, where

$C_j \in \{1, \dots, K\}$ for $j = 1, \dots, J$. We will refer to the values in $\boldsymbol{C}$ as column cluster (CC) labels. Then, we consider a set of membership labels for the rows *within every possible column cluster*, defining $\boldsymbol{R}_k = \left(R_{1,k}, \dots, R_{N,k}\right)$, with $R_{i,k} \in \{1, \dots, L\}$ for all $i = 1, \dots, N$ and $k = 1, \dots, K$. Following [20], the basic model specification under the separately exchangeable setting is the following:

$$
\begin{aligned}
p(\boldsymbol{Y} \mid \boldsymbol{R}, \boldsymbol{C}, \boldsymbol{\theta}) &= \prod_{i=1}^{N} \prod_{j=1}^{J} f(y_{i,j} \mid \theta_{R_{i,C_j}}) \\
p(\boldsymbol{R} \mid \boldsymbol{\omega}) &= \prod_{i=1}^{N} \prod_{k=1}^{K} \omega_{R_{i,k},k} \\
p(\boldsymbol{C} \mid \boldsymbol{\pi}) &= \prod_{j=1}^{J} \pi_{C_j}
\end{aligned}
\tag{1}
$$

where $f(\cdot|\theta)$ is a suitable parametric density for the data. Namely, the matrix entries $y_{i,j}$ associated to the same observational cluster l, i.e., the observations belonging to the set $\{y_{i,j} : R_{i,C_j} = l\}$, are modeled as conditionally i.i.d. from $f(\cdot|\theta_l)$. Also, the labels $\boldsymbol{R}$ and $\boldsymbol{C}$ are modeled with categorical random variables, such that $\mathbb{P}\left[R_{i,k} = l|\boldsymbol{\omega}\right] = \omega_{l,k}$ and $\mathbb{P}\left[C_j = k|\boldsymbol{\pi}\right] = \pi_k$. Essentially, we cluster together different *m/z* if they have similar activation levels (a trait captured by the kernel parameters $\{\theta_l\}_{l=1}^{L}$), and we group together different pixels if they have similar activation profiles, as represented by the vector of cluster assignment of the *m/z*. Note that according to the specification in (1), the distributional law of the data matrix $\boldsymbol{Y}$ is invariant to permutations of entire rows and columns. Moreover, given the nested nature of the model, the clustering of the signals is affected by the CCs, and the cluster assignment of a given signal is the same for all the pixels belonging to the same CC.

For our application, we first transform the original abundance matrix, denoted with $\boldsymbol{Z}$, into real values as

$$
y_{i,j} = \Phi^{-1}\left(\frac{z_{i,j} - \min_{i,j}(z_{i,j})}{\max_{i,j}(z_{i,j}) - \min_{i,j}(z_{i,j})}\right), \quad \text{for} \quad i = 1, \dots, N; \quad j = 1, \dots, J,
$$

where $\Phi^{-1}(\cdot)$ represents the inverse c.d.f. of a Normal distribution. Then, we assume that each mixture kernel composing the likelihood in (1) is a Normal density and specify a conjugate Normal-Inverse Gamma prior on its parameters $\theta_l = \left(\mu_l, \sigma_l^2\right)$, for $l = 1, \dots, L$. This implies that $\mu_l \mid \sigma_l^2 \sim N(m_l, \sigma_l^2/\kappa_l)$ and $\sigma_l^2 \sim IG(a_l, b_l)$ where we set $m_l = 0$, $\kappa_l = 0.01$, $a_l = 3$, and $b_l = 2$. The model is then completed with prior specifications for the cluster weights $\boldsymbol{\omega}_k$ and $\boldsymbol{\pi}$. In [20], the authors consider a fully nonparametric approach, letting both L and K be infinite and adopting stick-breaking Dirichlet process representations for their respective weights. Here, we consider a finite version of the model, following [10, 11], and adopt sparse Dirichlet distributions for the mixture weights. In other words, we set $\boldsymbol{\omega}_k \sim Dir_L(\boldsymbol{b}_0)$ for $k = 1, \ldots, K$ and $\boldsymbol{\pi} \sim Dir_K(\boldsymbol{a}_0)$ with $\boldsymbol{a}_0, \boldsymbol{b}_0$ vectors identically equal to 0.0001.

We perform posterior inference using a tailored variational Bayes (VB) algorithm to grant an efficient estimation of this nested model on the high-dimensional MALDI-MSI data. In short, let $\boldsymbol{\Theta} = (\boldsymbol{R}, \boldsymbol{C}, \boldsymbol{\theta}, \boldsymbol{\pi}, \boldsymbol{\omega})$ represent the set of all the parameters in our model. After a family of approximating densities $\mathcal{Q}$ is specified over $\boldsymbol{\Theta}$, we seek to find the member $q^*(\boldsymbol{\Theta})$ of $\mathcal{Q}$ such that

$$q^*(\boldsymbol{\Theta}) = \arg\min_{q_\Lambda \in \mathcal{Q}} \mathcal{D}_{KL}\left(q_\Lambda(\boldsymbol{\Theta}) \mid\mid p(\boldsymbol{\Theta} \mid \boldsymbol{Y})\right), \tag{2}$$

i.e., we look for the candidate that minimizes the Kullback-Leibler (KL) divergence with true posterior distribution. Note that minimizing the KL divergence in (2) is equivalent to maximizing the Evidence Lower BOund (ELBO), defined as $ELBO(q) = \mathbb{E}\left[\log p\left(\boldsymbol{\Theta}, \boldsymbol{Y}\right)\right] - \mathbb{E}\left[\log q\left(\boldsymbol{\Theta}\right)\right]$. We opted for a mean-field variational approximation, which chooses $\mathcal{Q}$ assuming the parameters are mutually independent. For an introduction to VB methods, the reader is referred to Chap. 10 of [4] and [5].

3 Biclustering Results

For the MALDI-MSI data analysis, we set the upper bounds for the numbers of clusters across columns and across rows to $K = 30$ and $L = 40$, respectively. We ran our model 100 times using different starting points and retained the estimate that obtained the highest value of the ELBO. Once the algorithm has converged, we are left with two sets of posterior variational clustering probabilities that we denote with $\hat{\rho}_{j,k} = q^*(C_j = k)$ and $\hat{\xi}_{i,l,k} = q^*(R_{i,k} = l)$. The former indicates how likely the j-th column is to be assigned to the k-th cluster. Similarly, the latter represents the probability of the i-th row to belong to the l-th row cluster within the k-th CC. Thus, we estimate the pixels segmentation as

$$\hat{C}_j = \arg\max_{k=1,\ldots,K} \hat{\rho}_{j,k} \;\; \forall j. \tag{3}$$

Analogously, the row-clustering solutions nested in the k-th column cluster are given by

$$\hat{R}_{i,k} = \arg\max_{l=1,\ldots,L} \hat{\xi}_{i,k,l} \;\; \forall i, k. \tag{4}$$

It is important to note that, with this procedure, we obtain K partitions of the rows, one for each *potential* CC. However, only the ones corresponding to non-empty CCs are interesting. In our application, the number of estimated CCs is 25.

The left panel of Fig. 2 displays the retrieved image segmentation. The clusters are colored according to the average transformed abundance level, ranging from dark red

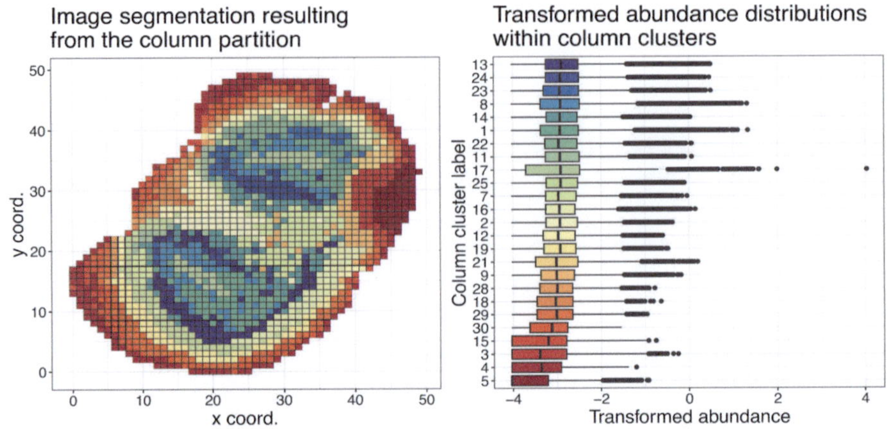

Fig. 2 The left panel shows a heatmap indicating the clustering assignments obtained via our VB estimation of the model. The right panel reports the associated boxplots of signals assigned to each CC. The colors are ordered from dark red to dark blue to reflect the average abundance of the clusters (estimated number of CCs: 25)

(lowest average abundance) to dark blue (highest average abundance). Despite the simplicity of the proposed model, clustering according to profile differences allows us to recover a good separation between the inter-brain formations (hippocampal formation and white matter regions, which are also highlighted in the bottom left panel of Fig. 3) and the external ones (gray matter, featured in the top left panel of Fig. 3) in line with the findings of [9, 13]. Moreover, the estimate clearly captures the sample malformation due to technical challenges in the sample preparation that we mentioned in Sect. 1. From a biological perspective, there is a high degree of similarity between the recovered segmentation and the recognized brain regions of the Mouse Brain Atlas from the Allen Institute [14, 23].

In the right panel of Fig. 2 we summarize the signal activations in each CC by reporting a collection of boxplots that describe the distribution of the transformed abundances across the CCs. The boxplots are colored to reflect the segmentation in the left panel, and ordered according to their averages. It is reassuring to recover that on the one hand, inter-brain formations are associated with higher signal values, as expected. On the other hand, the outer brain part and the problematic area are characterized by lower levels of activation. Moreover, our model detects that the right side of the slide is characterized by unusually low abundance, and it separates it from the rest of the outer border, as it is also highlighted in the central left panel of Fig. 3.

Note that if we were only interested in segmenting the distributions of the abundances in each pixel, then a partially exchangeable model would have been sufficient. Instead, our separately exchangeable model allows us to recover even deeper insights. Indeed, our model allows us to study the activation profiles across the CCs, i.e., the differences in signals across different estimated brain regions. Three examples are

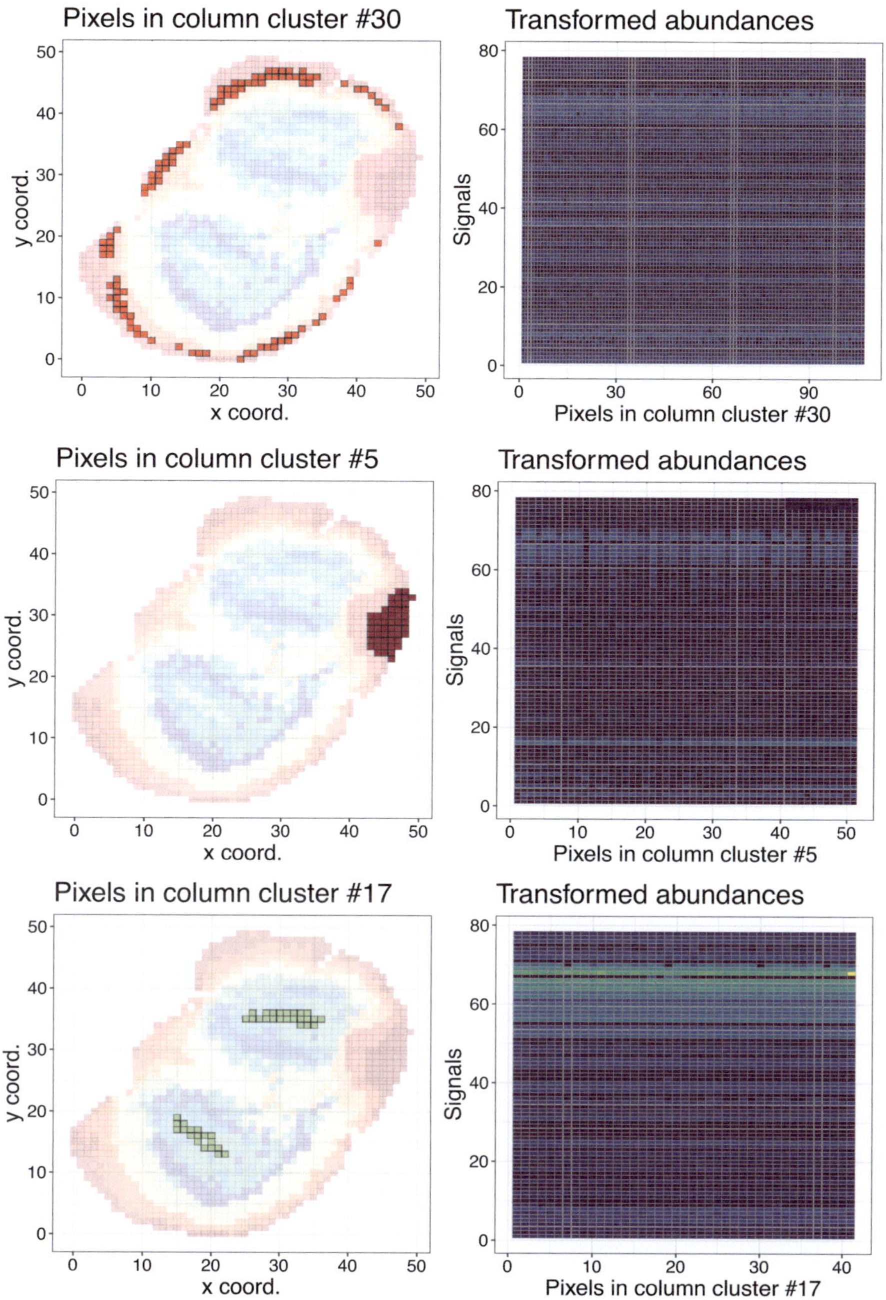

Fig. 3 Focus on CCs related to three important brain regions: gray matter (top row), problematic area (central row), and inter-brain formations (bottom row). The panels on the left highlight the recovered regions. The heatmaps on the right show the abundance profiles for the corresponding columns (pixels). Different activation patterns across signals are evident

displayed in Fig. 3, where we report part of the peripheral region (top row), the problematic area (central row), and inter-brain formations (bottom row). The panels on the left highlight the regions recovered by each CCs. The heatmaps on the right show the abundance profiles across all signals for the pixels corresponding to the selected brain region. Different activation patterns across signals are evident, highlighting the putative molecular markers of specific anatomically distinct brain regions. These results pave the way for further biological analyses that could demonstrate the connectivity between the found molecule classes with different physiochemical properties.

While not visually highlighted in Fig. 3, a different partition of the rows is available for each estimated brain region, and the benefit of our model comes from the interaction between the CCs and the partition of the signals. Recall that each row cluster is associated with one of the kernel parameters θ_l's. By analyzing the associated θ_l's, for example, one could retrieve the most under- and over-expressed signals in each CC by focusing on the row clusters corresponding to the lowest and highest values of the posterior expected values of μ_l, respectively. This could be especially beneficial to summarize and reduce the dimensionality of the recovered findings, in order to communicate the results to clinicians, who can then identify the biological markers of each identified region.

4 Conclusions and Future Directions

We have reported some preliminary results that prove the effectiveness of using separately exchangeable models estimated via VB on high-dimensional mass spectrometry imaging data. There are several future directions of interest to improve this model. First, we aim at incorporating the spatial information into the model using Hidden Markov Random fields [HRMF, 3, 18]. Second, Bayesian nonparametric priors can be combined with the HMRF specification to obtain a flexible distribution for the cluster weights and avoid the ex-ante specification of the number of clusters, as suggested in [17]. Finally, we are also considering multiomics avenues, developing a method to incorporate information from molecules other than lipids, such as N-glycans and peptides.

References

1. Balocchi, C., George, E.I., Jensen, S.T.: Clustering Areal Units at Multiple Levels of Resolution to Model Crime in Philadelphia (2021). http://arxiv.org/abs/2112.02059
2. Beraha, M., Guglielmi, A., Quintana, F.A.: The semi-hierarchical Dirichlet process and its application to clustering homogeneous distributions. Bayesian Anal. **16**(4), 1187–1219 (2021). https://doi.org/10.1214/21-BA1278
3. Besag, J.: On the statistical analysis of dirty pictures. J. Roy. Stat. Soc. Ser. B (Methodological) **48**(3), 259–302 (1986)
4. Bishop, C.M.: Pattern Recognition and Machine Learning. Springer, New York, NY (2006)

5. Blei, D.M., Kucukelbir, A., McAuliffe, J.D.: Variational inference: a review for statisticians. J. Am. Stat. Assoc. **112**(518), 859–877 (2017). https://doi.org/10.1080/01621459.2017.1285773
6. Boggio, K., Obasuyi, E., Sugino, K., Nelson, S., Agar, N., Agar, J.: Recent advances in single-cell MALDI mass spectrometry imaging and potential clinical impact. Expert Rev. Proteomics **8**(5), 591–604 (2011). https://doi.org/10.1586/epr.11.53
7. Camerlenghi, F., Dunson, D.B., Lijoi, A., Prünster, I., Rodríguez, A.: Latent nested nonparametric priors (with discussion). Bayesian Anal. **14**(4) (2019). https://doi.org/10.1214/19-ba1169. http://arxiv.org/abs/1801.05048
8. Capitoli, G., van Abeelen, K., Piga, I., L'Imperio, V., Besozzi, D., Nobile, M., Galimberti, S.: Well begun is half done: the impact of pre-processing in MALDI mass spectrometry analysis. An application to thyroid cancer diagnosis. Working Paper (2023)
9. Capitoli, G., Colombara, S., Cotroneo, A., De Caro, F., Morandi, R., Schembri, C., Zapiola, A., Denti, F.: Detecting latent spatial patterns in mass spectrometry brain imaging data via Bayesian mixtures. In: Pearson (ed.) Book of Short Papers SIS 2023, pp. 1127–1132. Pearson (2023)
10. D'Angelo, L., Canale, A., Yu, Z., Guindani, M.: Bayesian nonparametric analysis for the detection of spikes in noisy calcium imaging data. Biometrics (2022). https://doi.org/10.1111/biom.13626
11. D'Angelo, L., Denti, F.: A finite-infinite shared atoms nested model for the Bayesian analysis of large grouped data. Bayesian Anal. (2024). https://arxiv.org/html/2406.13310v1
12. Denti, F., Camerlenghi, F., Guindani, M., Mira, A.: A common atoms model for the Bayesian nonparametric analysis of nested data. J. Am. Stat. Assoc. 1–12 (2021). https://doi.org/10.1080/01621459.2021.1933499
13. Denti, V., Capitoli, G., Piga, I., Clerici, F., Pagani, L., Criscuolo, L., Bindi, G., Principi, L., Chinello, C., Paglia, G., et al.: Spatial multiomics of lipids, n-glycans, and tryptic peptides on a single FFPE tissue section. J. Proteome Res. **21**(11), 2798–2809 (2022)
14. Dong, H.W.: The Allen Reference Atlas: A Digital Color Brain Atlas of the C57Bl/6J Male Mouse. John Wiley & Sons Inc (2008)
15. Lee, J., Müller, P., Zhu, Y., Ji, Y.: A nonparametric Bayesian model for local clustering with application to proteomics. J. Am. Stat. Assoc. **108**(503), 775–788 (2013). https://doi.org/10.1080/01621459.2013.784705
16. Lijoi, A., Prünster, I., Rebaudo, G.: Flexible clustering via hidden hierarchical Dirichlet priors. Scand. J. Stat. **50**(1), 213–234 (2023). https://doi.org/10.1111/sjos.12578
17. Lü, H., Arbel, J., Forbes, F.: Bayesian nonparametric priors for hidden Markov random fields. Stat. Comput. **30**(4), 1015–1035 (2020). https://doi.org/10.1007/s11222-020-09935-9
18. McGrory, C.A., Titterington, D.M., Reeves, R., Pettitt, A.N.: Variational Bayes for estimating the parameters of a hidden Potts model. Stat. Comput. **19**(3), 329–340 (2009). https://doi.org/10.1007/s11222-008-9095-6
19. Moran, G.E., Ročková, V., George, E.I.: Spike-and-Slab Lasso biclustering. Ann. Appl. Stat. **15**(1), 148–173 (2021). https://doi.org/10.1214/20-AOAS1385
20. Rebaudo, G., Lin, Q., Mueller, P.: Separate exchangeability as modeling principle in Bayesian nonparametrics (2021). http://arxiv.org/abs/2112.07755
21. Rodríguez, A., Dunson, D.B., Gelfand, A.E.: The nested Dirichlet process. J. Am. Stat. Assoc. **103**(483), 1131–1154 (2008). https://doi.org/10.1198/016214508000000553
22. Rohner, T., Staab, D., Stoeckli, M.: Maldi mass spectrometric imaging of biological tissue sections. Mech. Ageing Dev. **126**(1), 177–185 (2005). https://doi.org/10.1016/j.mad.2004.09.032
23. Sunkin, S.M., Ng, L., Lau, C., Dolbeare, T., Gilbert, T.L., Thompson, C.L., Hawrylycz, M., Dang, C.: Allen brain atlas: an integrated spatio-temporal portal for exploring the central nervous system. Nucleic Acids Res. **41**(D1), D996–D1008 (2012)

A Bayesian Approach with Gaussian Priors to the Inverse Problem of Source Identification in Elliptic PDEs

Matteo Giordano

Abstract We consider the statistical linear inverse problem of making inference on an unknown source function in an elliptic partial differential equation from noisy observations of its solution. We employ nonparametric Bayesian procedures based on Gaussian priors, leading to convenient conjugate formulae for posterior inference. We review recent results providing theoretical guarantees on the quality of the resulting posterior-based estimation and uncertainty quantification, and we discuss the application of the theory to the important classes of Gaussian series priors defined on the Dirichlet-Laplacian eigenbasis and Matérn process priors. We provide an implementation of posterior inference for both classes of priors, and investigate its performance in a numerical simulation study. The reproducible code is available at: https://github.com/MattGiord/Bayesian-Source-Identification.

1 Introduction

Linear inverse problems consist in the task of recovering unknown objects or physical quantities from linear indirect noisy measurements. A widespread mathematical formulation for these problems postulates that the recovery target be an element f of a Hilbert space $\mathbb{H}_1$, and that the data arise according to the equation

$$Y^\varepsilon = G(f) + \varepsilon W, \tag{1}$$

where $G : \mathbb{H}_1 \to \mathbb{H}_2$ is a linear operator between $\mathbb{H}_1$ and another Hilbert space $\mathbb{H}_2$, W is additive observational noise and $\varepsilon > 0$ is the noise level. In view of the central limit theorem, normality of the measurement errors can often be maintained, whereby W is assumed to be a white noise process indexed by $\mathbb{H}_2$. The goal is then to estimate f from an observed realisation of Y^ε.

M. Giordano (✉)
ESOMAS Department, University of Turin, Turin, Italy
e-mail: matteo.giordano@unito.it

A. Avalos-Pacheco et al. (eds.), *New Trends in Bayesian Statistics*,
Springer Proceedings in Mathematics & Statistics 511,
https://doi.org/10.1007/978-3-031-99009-0_2

Observation schemes as in Eq. (1) are found in a variety of scientific fields and engineering applications, including medical imaging [9], geophysics [40], acoustics [12] and finance [7]. For example, Computerised Tomography (CT), a technique to obtain detailed images of the human body and information about the density variation of the tissues, is based on a mathematical model (related to the 'Radon transform') for the absorption of X-rays. Similar concepts underpin many other medical imagining techniques, such as Magnetic Resonance (MR) and Positron Emission Tomography (PET); see [9] for further details and references.

In many such applications, the unknown f may be characterised as a functional coefficient governing a partial differential equation (PDE), while the observed object $G(f)$ is the corresponding PDE solution. The 'forward operator' is then the coefficient-to-solution map $G : f \mapsto G(f)$. See the monograph [24] for an extensive overview on PDE-based inverse problems.

In the present article, we shall mostly focus on the representative example of 'source identification' in elliptic PDEs. Many of the ideas developed hereafter have a natural application to other linear inverse problems. Let $\mathcal{O} \subset \mathbb{R}^d$ be an open and bounded set with smooth boundary $\partial\mathcal{O}$. Let the unknown (square-integrable) function $f \in \mathbb{H}_1 \equiv L^2(\mathcal{O})$ be the 'source term' in the elliptic PDE with zero Dirichlet boundary conditions,

$$\begin{aligned} \nabla \cdot (c\nabla u) - f &= 0, \quad \text{on } \mathcal{O}, \\ u &= 0, \quad \text{on } \partial\mathcal{O}, \end{aligned} \tag{2}$$

where $\nabla\cdot$ and ∇ denote, respectively, the divergence and gradient operators, and where the smooth and positive 'diffusion coefficient' $c \in C^\infty(\overline{\mathcal{O}})$, $\inf_{x\in\mathcal{O}} c(x) > 0$, is assumed to be known. By standard elliptic theory, e.g. [33, Chap. 2] and [14, Chap. 6], for any $f \in L^2(\mathcal{O})$ there exists a unique (weak) solution $G(f) \equiv u$ in the Sobolev space $H^1(\mathcal{O}) \subset L^2(\mathcal{O}) \equiv \mathbb{H}_2$, giving rise to a linear (injective, self-adjoint and compact) operator $G : L^2(\mathcal{O}) \to L^2(\mathcal{O})$; see Appendix B in [18] for further details. We then consider the problem of estimating the source f from observations Y^ε arising as in (1), with G the solution map associated to the PDE (2) and W a Gaussian white noise process indexed by $L^2(\mathcal{O})$. An illustration of the problem with synthetic data is provided in Fig. 1. Among the numerous applications areas, inverse problems based on elliptic PDEs are important building blocks in oil reservoir modelling [47]. Source identification problems have been extensively investigated in the applied mathematics and statistics communities; see [2, 13, 18, 21] and the many references therein.

Here, we shall pursue the popular nonparametric Bayesian approach to inverse problems [41]. We shall assign to the unknown source f a prior distribution $\Pi(\cdot)$ supported on the function space $L^2(\mathcal{O})$ and then, following the Bayesian paradigm, combine it with the (Gaussian) data likelihood induced by the statistical model (1) to form the posterior distribution $\Pi(\cdot|Y^\varepsilon)$ of $f|Y^\varepsilon$, representing our updated belief about the inferential target and providing us with point estimates and uncertainty quantification. In particular, Gaussian priors represent a natural choice for inference

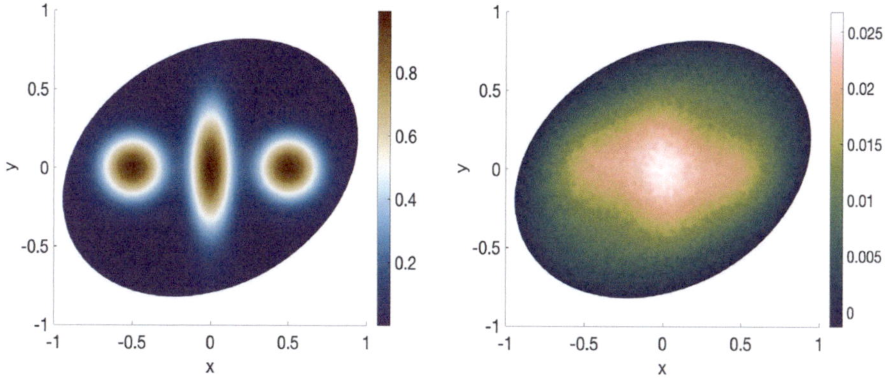

Fig. 1 Left: a source function f on a rotated elliptically-shaped domain, modelling three heat sources centred at the points $(-0.5, 0)$, $(0, 0)$ and $(0, 0.5)$. Right: noisy observations of the associated PDE solution $G(f)$ (with fixed diffusivity c)

in the context of the observation scheme (1) due to their conjugacy, which leads to convenient explicit formulae for the posterior distribution.

The Bayesian approach to inverse problems dates back at least to the 1980s, with early seminal work laid out in [31, 32, 43] among the others, and has since gained enormous popularity across several applied fields. See the monographs [25, 42], as well as the more recent reviews [6, 8], where further references can be found. Over the last decade, a large number of articles have investigated the theoretical recovery performance of Gaussian priors in linear inverse problems. We refer in particular to [4, 20, 27, 29, 30], and also mention [1, 17, 19, 35, 36] for results in nonlinear problems. Recently, Giordano and Kekkonen [18], building on earlier findings by [34], identified general conditions on the forward operator, on the ground truth and on the prior distribution under which 'semiparametric' Bernstein-von Mises theorems can be obtained, characterising the asymptotic shape of the posterior distribution of a large collection of linear functionals of the unknown. For the problem of estimating the source f in the elliptic PDE (2) from observations Y^ε arising as in (1), they showed that a wide class of 'standard' centred Gaussian process priors (such as the one associated to the commonly used Matérn kernel) yield, in the small noise limit $\varepsilon \to 0$ and under the frequentist assumption that the data have been generated by a fixed ground truth f_0, valid and optimal estimation and uncertainty quantification, via the posterior mean estimate and credible intervals centred around it, for one-dimensional aspects of the unknown source function.

For this study, we shall focus on the implementation of the Bayesian procedures with Gaussian priors for the inverse problem of source identification that were investigated in [18], corroborating the theory developed therein with a numerical simulation study. To this end, we will first review, in Sect. 2.2, the asymptotic results derived in [18]. We will then provide, in Sect. 2.3, two examples of Gaussian priors satisfying the assumptions of the general theory; in particular, we will consider centred

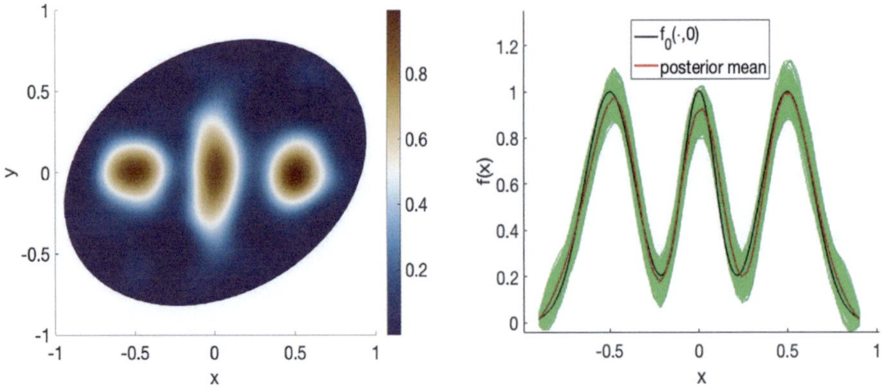

Fig. 2 Left: the posterior mean estimate of the source function f. Right: cross section along the x-axis of the true source function f_0 (in black), of the posterior mean (in red), and of 2500 posterior samples (in green)

Gaussian series priors defined on the eigenbasis of the Dirichlet-Laplacian, as well as centred stationary Gaussian process priors associated to the Matérn covariance kernel. Both classes of priors are of practical interest and widely used, and they will lead to two different discretisation strategies for the implementation.

We will present the numerical simulation study in Sect. 3, were we will provide an implementation of posterior inference in the source identification problem for Gaussian series priors and for the Matérn process priors. The series approach will hinge on a natural discretisation of the parameter space via high-dimensional basis expansions, while for the Matérn priors we will employ piecewise linear functions defined on the elements of a deterministic triangular mesh. Under a suitable discretisation of the statistical model (1), we will provide the explicit formulae for the Gaussian conjugate posterior distributions, which we will exploit to efficiently compute the posterior mean estimates and to efficiently implement, via posterior sampling, credible sets for uncertainty quantification. In the study, we will numerically investigate the asymptotic concentration of the posterior distribution around the ground truth generating the data, the efficiency of the posterior mean estimator (in terms of minimality of its asymptotic variance), and further the frequentist coverage of the obtained credible intervals. Overall, the study shows a close correspondence between the numerical results and the expected performance predicted by the theory developed in [18]. The posterior mean estimate (relative to a Gaussian series prior) and the associated uncertainty quantification (obtained via the cross-section of 2500 posterior samples along the x-axis) are depicted in Fig. 2, to be compared to the true source function shown in Fig. 1. The reproducible (MATLAB) code used for the study is available at: https://github.com/MattGiord/Bayesian-Source-Identification.

2 A Nonparametric Bayesian Approach with Gaussian Priors

In this section, we formally introduce the statistical model and the Bayesian procedures of interest for the present article. We will provide details on the likelihood, the prior and the posterior distribution in Sect. 2.1. We will then review the asymptotic results of [18] in Sect. 2.2, and further present two classes of Gaussian priors to which the theory applies in Sect. 2.3.

2.1 *Details on the Bayesian Model*

Throughout, let $\mathcal{O} \subset \mathbb{R}^d$, $d \in \mathbb{N}$, be a non-empty, open and bounded set with smooth boundary $\partial\mathcal{O}$. For a fixed, smooth and positive 'diffusion coefficient' $c \in C^\infty(\overline{\mathcal{O}})$, $\inf_{x\in\mathcal{O}} c(x) > 0$, let $G : L^2(\mathcal{O}) \to L^2(\mathcal{O})$ be the (linear) source-to-solution map associated to the elliptic PDE (2); see Appendix B in [18] for details. For such G, and a given noise level $\varepsilon > 0$, consider observations Y^ε arising as in (1) for some unknown $f \in L^2(\mathcal{O})$, where W is a white noise process indexed by $L^2(\mathcal{O})$ defined on some probability space $(\Omega, \mathcal{S}, \Pr)$, that is the centred Gaussian process $(W(g),\ g \in L^2(\mathcal{O}))$ with covariance $E[W(g_1)W(g_2)] = \langle g_1, g_2\rangle_2$. Throughout most of the article, we will regard ε as known. In practice, it can often be replaced by an estimate (cf. Sect. 3.3.2). As described, for example, in Chap. 1 of [16], observing Y^ε is understood as observing a realisation of the Gaussian process $(Y^\varepsilon(g),\ g \in L^2(\mathcal{O}))$ with mean $E[Y^\varepsilon(g)] = \langle G(f), g\rangle_2$ and covariance $\mathrm{Cov}[Y^\varepsilon(g_1), Y^\varepsilon(g_2)] = \langle g_1, g_2\rangle_2$. Such observation scheme serves as a convenient continuous counterpart of the inverse regression model

$$Y_i = G(f)(X_i) + \sigma W_i, \qquad i = 1, \dots, n, \tag{3}$$

comprising noisy point evaluations of the PDE solution $G(f)$ over a set of points $X_1, \dots, X_n \in \mathcal{O}$, corrupted by Gaussian measurement errors $\sigma W_1, \dots, \sigma W_n \overset{\text{iid}}{\sim} N(0, \sigma^2)$, $\sigma > 0$, known to be asymptotically equivalent (in the sense of [10]) to the white noise model (1) under suitable assumptions on the grid and the calibration $n/\sigma^2 \simeq \varepsilon^{-2}$.

For continuous observations Y^ε arising as in (1), for any $f \in L^2(\mathcal{O})$, the (cylindrically-defined) law P_f^ε of Y^ε is absolutely continuous with respect to the law P_0^ε of the scaled white noise εW, with log-likelihood

$$\ell_\varepsilon(f) := \log \frac{dP_f^\varepsilon}{dP_0^\varepsilon}(Y^\varepsilon) = \frac{1}{\varepsilon^2} Y^\varepsilon[G(f)] - \frac{1}{2\varepsilon^2}\|G(f)\|_2^2.$$

In view of the joint measurability of ℓ_ε, regarding f as a random function with values in $L^2(\mathcal{O})$ and assigning to it any prior distribution in the form of a Borel probability measure $\Pi(\cdot)$ supported on $L^2(\mathcal{O})$ then induces, via Bayes' formula (for example, [15, p. 7]), the posterior distribution

$$\Pi(A|Y^\varepsilon) = \frac{\int_A e^{\ell_\varepsilon(f)} d\Pi(f)}{\int_{L^2(\mathcal{O})} e^{\ell_\varepsilon(f')} d\Pi(f')}, \qquad A \subseteq L^2(\mathcal{O}) \text{ measurable},$$

that is, the conditional distribution of $f|Y^\varepsilon$. In particular, it will be of interest to consider Gaussian priors which, in view of the linearity of the forward operator G and the normal assumption on the noise W, will lead to conjugate Gaussian posteriors. Concrete formulae are provided in Sect. 3. In the following, we will repeatedly use elements of the theory of Gaussian processes and measures on Hilbert spaces, and we refer to [16, Chap. 2] for the necessary background. For a Gaussian prior $\Pi(\cdot)$ on $L^2(\mathcal{O})$, the 'information geometry' is encoded within an associated reproducing kernel Hilbert space (RKHS) of functions defined on the domain $\mathcal{O}$, strictly contained inside $L^2(\mathcal{O})$. Popular prior choices in applications and theoretical studies typically model functions belonging to a 'smoothness scale', with associated RKHS equal to (or included in) a Sobolev space $H^\alpha(\mathcal{O})$, for some regularity level $\alpha > 0$. These include Gaussian series priors defined on bases spanning the Sobolev scale, as well as stationary Gaussian processes with the Matérn covariance kernel; see [15, Chap. 11].

2.2 Theoretical Guarantees for Estimation and Uncertainty Quantification

The asymptotic properties of nonparametric Bayesian procedures with Gaussian priors in inverse problems have recently been investigated by Giordano and Kekkonen [18], resulting, under general 'regularity conditions' for the forward operator, for the ground truth and for the prior distribution, in semiparametric Bernstein-von Mises theorems that entail the convergence of certain one-dimensional posterior distributions to limiting Gaussian probability measures with minimal variance, in the small noise limit and under the frequentist assumption that the data have been generated by a fixed ground truth. These results were then leveraged in [18] to prove the asymptotic efficiency of the associated posterior mean estimators, as well as to derive theoretical guarantees certifying that credible intervals centred around them are asymptotically valid confidence intervals with minimal width. In this section, we provide a review of the findings of [18] for the inverse problem of source identification. These will later be corroborated by the results of the numerical simulation study presented in Sect. 3.

The investigation of [18] builds on the semiparametric approach to the Bernstein-von Mises theorem in infinite-dimensional statistical models developed by Castillo and Nickl [11] and later refined by Monard et al. [34] in the inverse problem setting. It is based on the study of the posterior distributions of a class of scaled and centred one-dimensional functionals of the unknown which, in the context of the source identification problem, take the form $\varepsilon^{-1}\langle f - \bar{f}_\varepsilon, \psi\rangle_2$ for test functions $\psi \in L^2(\mathcal{O})$, where $\bar{f}_\varepsilon := E^\Pi[f|Y^\varepsilon]$ is the posterior mean. Let $\mathcal{L}(\varepsilon^{-1}\langle f - \bar{f}_\varepsilon, \psi\rangle_2|Y^\varepsilon)$ denote the associated scaled and centred posterior distribution.

Theorem 1 (Theorem 4.1 in [18]) *Let $\Pi(\cdot)$ be a centred Gaussian Borel probability measure supported on $L^2(\mathcal{O})$ with RKHS equal to $H^\alpha(\mathcal{O})$ for some $\alpha > d/2$. Let $f_0 \in H^\beta(\mathcal{O})$, for some $\beta > \alpha - d/2$, be compactly supported inside $\mathcal{O}$, and consider observations $Y^\varepsilon \sim P^\varepsilon_{f_0}$ from the statistical model* (1) *with $G(f)$ the solution to the PDE* (2) *and $f = f_0$. Then, for any $\gamma > 2 + d/2$ and any compactly supported test function $\psi \in H^\gamma(\mathcal{O})$, we have*

$$\mathcal{L}(\varepsilon^{-1}\langle f - \bar{f}_\varepsilon, \psi\rangle_2|Y^\varepsilon) \xrightarrow{\mathcal{L}} N(0, \|\nabla\cdot(c\nabla\psi)\|_2^2), \tag{4}$$

in $P^\varepsilon_{f_0}$-probability as $\varepsilon \to 0$.

The result asserts that the random (data-dependent) one-dimensional probability distribution $\mathcal{L}(\varepsilon^{-1}\langle f - \bar{f}_\varepsilon, \psi\rangle_2|Y^\varepsilon)$ converges (in the topology of weak convergence) in probability to a centred normal distribution with variance $\|\nabla\cdot(c\nabla\psi)\|_2^2$. The latter can be shown to be minimal [18, Remark 2.4], as it coincides with the Cramér-Rao lower bound for estimating the one-dimensional quantity $\langle f, \psi\rangle_2$ from data Y^ε arising as in (1). Furthermore, the class of test functions ψ for which the convergence (4) is obtained is to be understood to be maximal, in the sense that the infinite-dimensional Gaussian probability measure with marginals identified by the right hand side of (4) is tight (a necessary condition for weak convergence) only when $\gamma > 2 + d/2$; see Lemma 4.2 in [18] and the related discussion.

A first important consequence of Theorem 1 is a central limit for the 'plug-in' posterior mean estimators $\langle \bar{f}_\varepsilon, \psi\rangle_2$ of the one dimensional aspects $\langle f, \psi\rangle_2$ of the unknown. Note that, for Gaussian priors, these can be efficiently computed via the explicit formulae for the conjugate Gaussian posteriors. The central limit follows, as argued in Remark 2.4 in [18], from the convergence of moments in the limit (4). In particular, under the assumptions of Theorem 1, it holds that

$$\varepsilon^{-1}\left(\langle \bar{f}_\varepsilon, \psi\rangle_2 - \langle f_0, \psi\rangle_2\right) \xrightarrow{d} N(0, \|\nabla\cdot(c\nabla\psi)\|_2^2), \tag{5}$$

as $\varepsilon \to 0$. In view of the aforementioned minimality of the asymptotic variance $\|\nabla\cdot(c\nabla\psi)\|_2^2$, the result indeed implies the asymptotic efficiency of the plug-in estimators $\langle \bar{f}_\varepsilon, \psi\rangle_2$.

The second implication of the Berstein-von Mises result stated in Theorem 1 concerns the coverage and width of credible intervals built around the efficient estimators $\langle \bar{f}_\varepsilon, \psi\rangle_2$, which can be shown to be asymptotically valid frequentist confidence intervals and to have diameter shrinking at the optimal parametric rate ε^{-1}. For any level $a \in (0, 1)$, consider the $(1 - a)\%$-credible interval

$$C_{\varepsilon,a} := \{z \in \mathbb{R} : |z - \langle \bar{f}_\varepsilon, \psi\rangle_2| \leq R_{\varepsilon,a}\}, \tag{6}$$

where $R_{\varepsilon,a} > 0$ is the $(1 - a/2)\%$-quantile of the one-dimensional (Gaussian) posterior distribution of $\langle f, \psi\rangle_2 | Y^\varepsilon$, so that

$$\Pi\left(f : \langle f, \psi\rangle_2 \in C_{\varepsilon,a} | Y^\varepsilon\right) = 1 - a.$$

Then, in the setting of Theorem 1, the asymptotic frequentist coverage of $C_{\varepsilon,a}$ is given by

$$P^\varepsilon_{f_0}\left(\langle f_0, \psi\rangle_2 \in C_{\varepsilon,a}\right) \to 1 - a, \tag{7}$$

as $\varepsilon \to 0$, while its radius $R_{\varepsilon,a}$ satisfies

$$R_{\varepsilon,a} = O_{P^\varepsilon_{f_0}}(\varepsilon^{-1}).$$

See Corollary 2.5 in [18]. Note that although an analytic formulation of the credible intervals $C_{\varepsilon,a}$ requires the derivation of the quantiles of the one dimensional posterior distributions $\langle f, \psi\rangle_2 | Y^\varepsilon$, these can typically be numerically approximated by efficiently sampling from the explicitly available conjugate posterior distributions.

2.3 Examples of Gaussian Priors

In this section, we provide two concrete examples of Gaussian priors to which Theorem 1 applies. For both instances, an implementation of the resulting posterior inference will be presented in Sect. 3, based on two different discretisation strategies. The first example concerns Gaussian series priors.

Example 1 (*Gaussian series priors on the Dirichlet-Laplacian eigenbasis*) Let $(\phi_j,\ j \in \mathbb{N}) \subset H^1(\mathcal{O}) \cap C^\infty(\overline{\mathcal{O}})$ be the orthonormal basis of the space $L^2(\mathcal{O})$ formed by the eigenfunctions of the (negative) Dirichlet-Laplacian,

$$\begin{aligned} -\Delta\phi_j - \lambda_j\phi_j &= 0, \quad \text{on } \mathcal{O} \\ \phi_j &= 0, \quad \text{on } \partial\mathcal{O}, \end{aligned} \qquad j \in \mathbb{N}, \tag{8}$$

with associated eigenvalues $0 < \lambda_1 < \lambda_2 \le \lambda_3 \le \ldots$, satisfying $\lambda_j \to \infty$ as $j \to \infty$ according to Weyl's asymptotics, namely $\lambda_j = O(j^{2/d})$ as $j \to \infty$. We refer to Example 6.3 and Sect. 7.4 in [23] for details. The associated Hilbert scale

$$\mathbb{H}^\alpha := \left\{ f \in L^2(\mathcal{O}) : \|f\|^2_{\mathbb{H}^\alpha} := \sum_{j=1}^\infty \lambda_j^\alpha |\langle f, \phi_j\rangle_2|^2 < \infty \right\}, \qquad \alpha \ge 0,$$

then satisfies $\mathbb{H}^0 = L^2(\mathcal{O})$ (with equality of norms) and the continuous (strict) embedding $\mathbb{H}^\alpha \subset H^\alpha(\mathcal{O})$ for all $\alpha > 0$ [44, p. 472]. In fact, it holds that $\|f\|_{\mathbb{H}^\alpha} \simeq \|f\|_{H^\alpha}$ for all $f \in \mathbb{H}^\alpha$ and $\alpha \ge 0$.

For any $\alpha > d/2$, consider the Gaussian random series

$$F := \sum_{j=1}^{\infty} \lambda_j^{-\alpha/2} F_j \phi_j, \qquad F_j \overset{\text{iid}}{\sim} N(0, 1), \tag{9}$$

corresponding to the Karhunen-Loève expansions of certain commonly used Gaussian process priors with covariance kernel given by an inverse power of the Laplacian [41, Sect. 2.4]. By Weyl's asymptotics, we have

$$\mathrm{E}[\|F\|_2^2] = \sum_{j=1}^{\infty} \lambda_j^{-\alpha} \simeq \sum_{j=1}^{\infty} j^{-2\alpha/d} < \infty,$$

since $2\alpha/d > 1$, showing that F takes values almost surely in $L^2(\mathcal{O})$. By Lemma I.5 in [15], the law $\Pi(\cdot)$ of F is then seen to define a Gaussian Borel probability measure supported on $L^2(\mathcal{O})$. Furthermore, by Theorem I.12 in [15], its RKHS is equal to $\mathbb{H}^\alpha$. Noting that, for any compactly supported test function $\psi \in H^\gamma(\mathcal{O})$, $\gamma > 2 + d/2$, the approximation argument in the proof of Theorem 4.1 in [18] can be carried out with minimal modifications with $H^\alpha(\mathcal{O})$ replaced by $\mathbb{H}^\alpha$, we conclude that Theorem 1 applies by modelling the unknown source function f via the Gaussian series (9).

While not explicitly available for general domains $\mathcal{O}$, we note that the Dirichlet-Laplacian eigenbasis can be numerically computed via efficient finite element methods for elliptic eigenvalue problems, offering a broadly applicable framework for implementation. More details will be provided in Sect. 3.

A second example of interest involves stationary Gaussian processes defined via a covariance kernel of choice. In particular, the Matérn kernel is widely used in applications [37, Sect. 4.2].

Example 2 (*Matérn process priors*) Let $F = (F(x),\ x \in \mathcal{O})$ be the centred and stationary Gaussian process with Matérn covariance kernel

$$C_{\alpha,\ell}(x, y) = \frac{2^{1-\alpha}}{\Gamma(\alpha)} \left(\frac{|x - y|\sqrt{2\alpha}}{\ell} \right)^{\alpha} B_\alpha \left(\frac{|x - y|\sqrt{2\alpha}}{\ell} \right), \qquad x, y \in \mathcal{O}, \tag{10}$$

with regularity parameter $\alpha > d/2$ and length scale $\ell > 0$. Above, Γ denotes the gamma function and B_α is the modified Bessel function of the second kind. The finite dimensional distributions of F are identified by the relation

$$(F(x_1), \dots, F(x_M))^T \sim N_M(0, \mathbf{C}), \tag{11}$$

with $\mathbf{C} := (C_{\alpha,\ell}(x_h, x_m))_{h,m=1}^M \in \mathbb{R}^{M,M}$, holding for any $M \in \mathbb{N}$ and any $x_1, \dots, x_M \in \mathcal{O}$. By Lemma I.4 in [15], a version of F can be identified with sample paths belonging almost surely to the Hölder space $C^{\alpha'}(\mathcal{O}) \subset L^2(\mathcal{O})$ for any $0 < \alpha' < \alpha - d/2$, and therefore, in view of Lemma I.7 in [15], the law $\Pi(\cdot)$ of such version defines a Gaussian Borel probability measure supported on $L^2(\mathcal{O})$. Moreover, the results in Sect. 11.4.4 in [15] imply that the RKHS of F equals, with norm equivalence, the set of restrictions to the domain $\mathcal{O}$ of functions in the Sobolev space $H^\alpha(\mathbb{R}^d)$. Since $\mathcal{O}$ is assumed to have smooth boundary, the latter is indeed equal to $H^\alpha(\mathcal{O})$. Thus, Theorem 1 applies with $\Pi(\cdot)$ a Matérn process prior with covariance kernel (10).

3 Numerical Simulation Study

For illustration, we take as working domain the area $\mathcal{O}$ contained inside a rotated ellipse with horizontal semi-axis of unit length, vertical semi-axis of length 3/4, and rotation angle $\theta = \pi/6$,

$$\{(\cos(t)\cos(\theta) - 3/4\sin(t)\sin(\theta), 3/4\sin(t)\cos(\theta) + \cos(t)\sin(\theta)),\ t \in [0, 2\pi)\}.$$

For an unknown source function $f \in L^2(\mathcal{O})$, we assume in practice that we are given n noisy point evaluations $\mathbf{Y} := (Y_1, \dots, Y_n)^T \in \mathbb{R}^n$ of the solution $G(f)$ to the PDE (2) generated according to the equivalent discrete statistical model (3), for a given deterministic grid of points $x_1, \dots, x_n \in \mathcal{O}$ comprising the nodes of a triangular mesh covering $\mathcal{O}$ (Fig. 3, top-left). We then seek to estimate f from data $\mathbf{Y}$.

3.1 Posterior Inference with Gaussian Series Priors

3.1.1 Methodology

For the Gaussian series priors defined via the Dirichlet-Laplacian eigenpairs $\{(\phi_j, \lambda_j),\ j \in \mathbb{N}\}$ considered in Example 1, we discretise the parameter space by modelling the unknown source function f as the finite sum

$$f = \sum_{j=1}^J f_j \phi_j, \qquad f_1, \dots, f_J \in \mathbb{R}, \qquad J \in \mathbb{N}. \tag{12}$$

For any such f, the linearity of the forward map G then implies that the discrete observations are given by

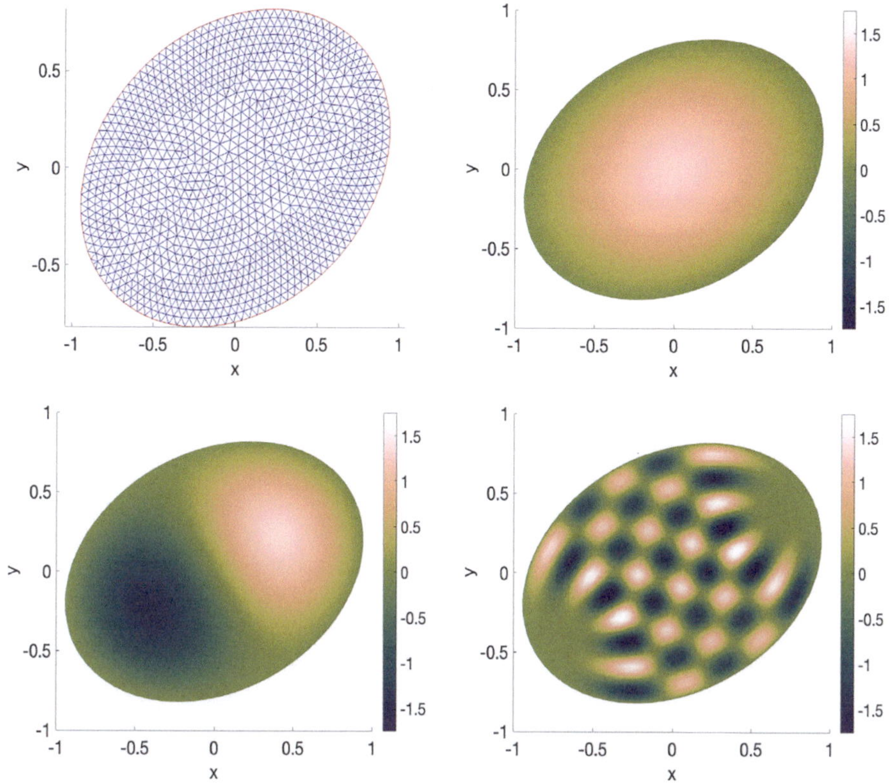

Fig. 3 Top-left: triangular mesh on a rotated elliptically-shaped domain with $n = 4500$ nodes. Top-right, bottom-left and bottom-right, respectively: numerical approximations of the first Dirichlet-Laplacian eigenfunction ϕ_1, of the second eigenfunction ϕ_2 and of the eigenfunction ϕ_{84} associated to the largest eigenvalue $\lambda_{84} = 493.3725$ found in the range $[0, 500]$

$$Y_i = \sum_{j=1}^{J} f_j G(\phi_j)(x_i) + \sigma W_i \equiv (\mathbf{Gf})_i + \sigma W_i$$

where $\mathbf{G} := [G(\phi_j)(x_i),\ \ i = 1, \dots, n,\ \ j = 1, \dots, J] \in \mathbb{R}^{n,J}$ and $\mathbf{f} := (f_1, \dots, f_J)^T \in \mathbb{R}^J$, whereby the inverse regression model (3) can be written in matrix notation as

$$\mathbf{Y} = \mathbf{Gf} + \sigma \mathbf{W} \tag{13}$$

with $\mathbf{W} := (W_1, \dots, W_n)^T \sim N_n(0, \mathbf{I}_n)$. Thus, for any given $\mathbf{f} \in \mathbb{R}^J$, $\mathbf{Y}|\mathbf{f} \sim N_n(\mathbf{Gf}, \sigma^2 \mathbf{I}_n)$.

In practice, outside certain special cases for the domain $\mathcal{O}$ (such as squared or circular ones), the Dirichlet-Laplacian eigenpairs are not explicitly available. For general domains, we then resort to numerical methods to solve the elliptic eigenvalue

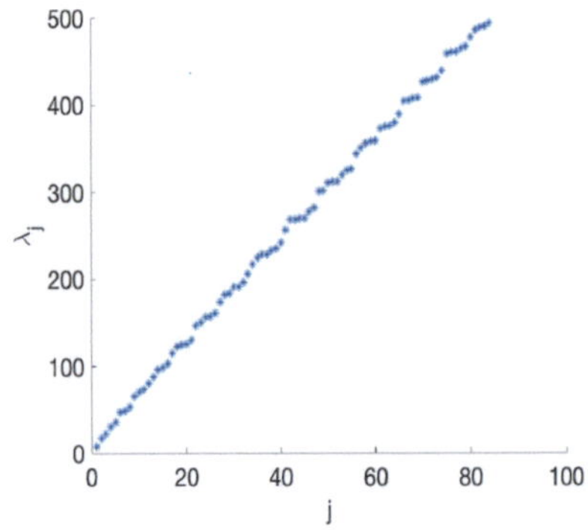

Fig. 4 Numerical approximations of the Dirichlet-Laplacian eigenvalues in the range $[0, 500]$

problem (8). In particular, we used MATLAB PDE Toolbox, that allows to input a range of search $[0, \lambda_{\max}]$, $\lambda_{\max} > 0$, for the eigenvalues, and then returns numerical approximations, obtained via finite element methods, of the eigenvalues in the prescribed interval and of the corresponding eigenfunctions. For the considered rotated elliptically-shaped domain, Fig. 3 shows the first, the second and the last eigenfunction returned by the elliptic PDE solver. The range was set to $[0, \lambda_{\max}] = [0, 500]$, for which $J = 84$ eigenvalues were found. The (numerical approximations to) the eigenvalues are displayed in Fig. 4. They exhibit a linear growth as expected from Weyl's asymptotics for bi-dimensional domains. The computation of the matrix $\mathbf{G}$ in the discretised observation model (13) is performed by numerically solving, again using MATLAB PDE Toolbox, the elliptic PDE (2) with f replaced by ϕ_j (or, more precisely, by the finite element approximation of ϕ_j) and then evaluating $\mathbf{G}_{ij} := G(\phi_j)(x_i)$ for all $i = 1, \dots, n$ and $j = 1, \dots, J$.

Under the discretisation (12), the Gaussian series prior described in Example 1 is approximately implemented by truncating the random series (9) at level J, and then assigning to the coefficients $f_1, \dots, f_J$ in (12) independent Gaussian priors $N(0, \lambda_j^{-\alpha})$, $j = 1, \dots, J$. In the discretised observation model (13), this corresponds to assigning to the vector $\mathbf{f}$ the J-dimensional Gaussian prior with diagonal covariance matrix

$$\mathbf{f} \sim N_J(0, \mathbf{\Lambda}), \qquad \mathbf{\Lambda} := \text{diag}(\lambda_1^{-\alpha}, \dots, \lambda_J^{-\alpha}) \in \mathbb{R}^{J,J}. \tag{14}$$

Thus, recalling that, according to (13), $\mathbf{Y}|\mathbf{f} \sim N_n(\mathbf{G}\mathbf{f}, \sigma^2\mathbf{I}_n)$, a standard conjugate computation for multivariate models with Gaussian likelihood and prior yields the posterior distribution

$$.\mathbf{f}|\mathbf{Y} \sim N_J(\bar{\mathbf{f}}_n, \mathbf{\Lambda}_n), \tag{15}$$

where

$$\bar{\mathbf{f}}_n := \frac{1}{\sigma^2}\mathbf{\Lambda}_n\mathbf{G}^T\mathbf{Y}; \qquad \mathbf{\Lambda}_n := (\sigma^{-2}\mathbf{G}^T\mathbf{G} + \mathbf{\Lambda}^{-1})^{-1}. \tag{16}$$

Using the conjugate formulae (15) and (16), it is straightforward to compute posterior mean estimates and drawing posterior samples. In turn, this allows to efficiently implement credible sets centred around the posterior mean, replacing the theoretical posterior quantiles (for example, the ones used in the definition of the credible intervals (6)) with the empirical quantiles associated to a sufficiently large sample from the posterior distribution.

3.1.2 Experiments

Throughout the numerical simulation study, the true source function (shown in Fig. 1, left) was taken to be

$$f_0(x, y) = e^{-(5x-2.5)^2-(5y)^2} + e^{-(7.5x)^2-(2.5y)^2} + e^{-(5x-2.5)^2-(5y)^2}, \qquad (x, y) \in \mathcal{O}.$$

Figure 1 (right) shows $n = 4500$ discrete noisy observations, over the nodes of the triangular mesh depicted in Fig. 3 (top-left), of the corresponding PDE solution $G(f_0)$ arising as in the inverse regression model (3) with noise standard deviation $\sigma = 0.0005$ (with corresponding signal-to-noise ratio $\|G(f_0)\|_2/\sigma = 37.55$). The diffusion coefficient was taken to be $c(x, y) := 2 + 5e^{-(5x-2)^2-(5y-2)^2} + 5e^{-(5x+2)^2-(5y+2)^2}$,
$(x, y) \in \mathcal{O}$. The PDE solution $G(f_0)$ was calculated using MATLAB PDE Toolbox, which also contains the routine to create the triangular mesh.

The posterior mean estimate $\bar{f}_n := \sum_{j=1}^{J} \bar{\mathbf{f}}_{n,j}\phi_j$ shown in Fig. 2 (left) was obtained by computing the vector of coefficients $\bar{\mathbf{f}}_n$ according to the conjugate formula in (16). A diagonal Gaussian prior as in (14) was used, with regularity parameter $\alpha = 3/4$. The parameter space was discretised using $J = 84$ basis functions. The obtained L^2-estimation error was $\|\bar{f}_n - f_0\|_2 = 0.060077$. For comparison, $\|f_0\|_2 = 0.4764$ (with corresponding relative error $\|\bar{f}_n - f_0\|_2/\|f_0\|_2 = 12.5\%$), while the L^2-approximation error incurred by projecting f_0 onto the linear space spanned by the employed set of basis functions (furnishing a lower bound for the L^2-estimation error) is 0.0486. The 2500 posterior draws whose cross-sections along the x-axis are shown in Fig. 2 (right) were sampled from the conjugate Gaussian posterior distribution in (15).

Figure 5 provides an illustration of asymptotic convergence in the infinitely-informative data limit, showing the posterior mean estimates obtained for increasing sample sizes. The (decreasing) L^2-estimation errors for sample sizes ranging between $n = 50$ and $n = 4500$ are reported in Table 1. Across the experiments, the same discretisation with $J = 84$ basis function and the same diagonal Gaussian prior with regularity $\alpha = 3/4$ were used.

We next consider semiparametric inference for one-dimensional linear functionals $\langle f, \psi \rangle_2$, $\psi \in L^2(\mathcal{O})$, and provide a numerical illustration of the asymptotic results presented in Sect. 2.2. In particular, we focus on test functions $\psi = \phi_j$, $j \in \{1, \dots, J\}$, belonging to the Dirichlet-Laplacian eigenbasis, for which, under

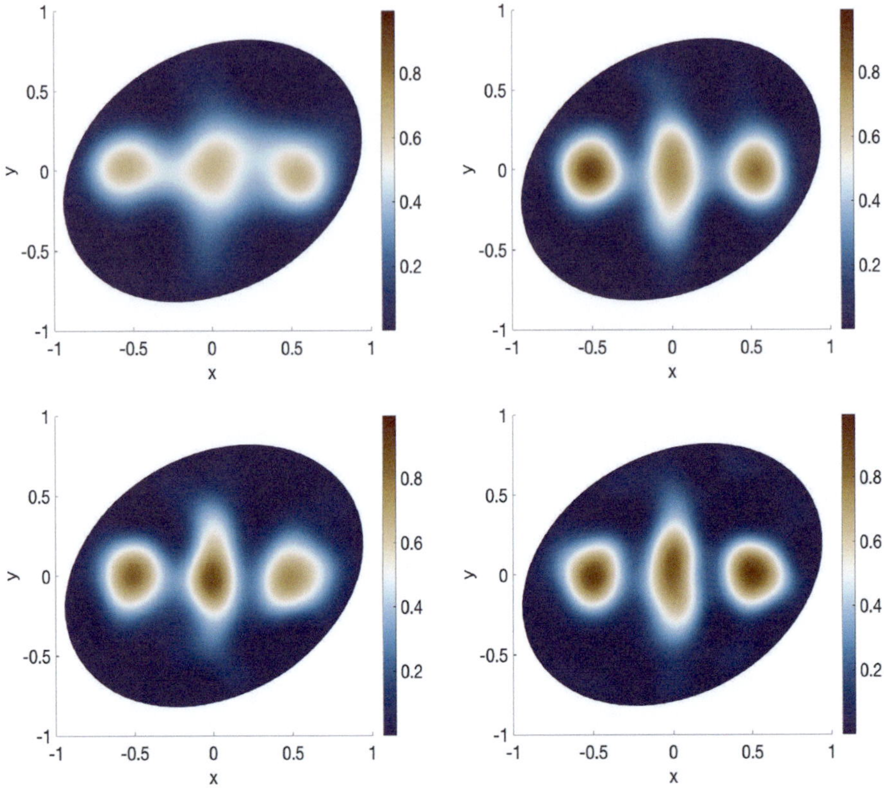

Fig. 5 Left to right, top to bottom: posterior mean estimates $\bar{f}_n$ of the source function f for increasing sample sizes $n = 100, 250, 500, 2000$

Table 1 L^2-estimation errors achieved by the posterior mean estimator $\bar{f}_n$ for increasing sample sizes

n	50	100	250	500	750	1000	2000	3000	4500
$\|\bar{f}_n - f_0\|_2$	0.22	0.18	0.13	0.099	0.088	0.078	0.076	0.069	0.060
$\|\bar{f}_n - f_0\|_2/\|f_0\|_2$	45.8%	37.5%	27.1%	20.6%	18.3%	16.3%	15.8%	14.3%	12.5%

the discretisation (12), $\langle f, \phi_j\rangle_2 = f_j$. Accordingly, for $\bar{\mathbf{f}}_n$ and $\mathbf{\Lambda}_n$ as in (16), the plug-in posterior estimators are given by $\langle \bar{f}_n, \phi_j\rangle_2 = \bar{\mathbf{f}}_{n,j}$, with corresponding posterior variances $\Gamma_{n,jj}$. Thus, the 95%-credible interval for $\langle f, \phi_j\rangle_2$ is given by $\bar{\mathbf{f}}_{n,j} \pm 1.96\sqrt{\Gamma_{n,jj}}$.

In order to compute the asymptotic variances $\|\nabla \cdot (c\nabla\psi)\|_2^2$ appearing in the right hand side of (4) and (5), we obtain the singular value decomposition (SVD) of the forward operator G, corresponding to finding the eigenfunctions $(\xi_k,\ k \in \mathbb{N}) \subset L^2(\mathcal{O})$ and eigenvalues $(\eta_k,\ k \in \mathbb{N}) \subset [0, \infty)$ solving the problem

$$\begin{aligned} -\nabla \cdot (c\nabla \xi) - \eta \xi &= 0, \quad \text{on } \mathcal{O} \\ \xi &= 0, \quad \text{on } \partial\mathcal{O}, \end{aligned} \tag{17}$$

whereupon there follow the identities

$$G(f) = \sum_{k=1}^{\infty} \eta_k^{-1} \langle f, \xi_k \rangle_2 \xi_k,$$

and

$$\nabla \cdot (c\nabla u) = \sum_{k=1}^{\infty} \eta_k \langle u, \xi_k \rangle_2 \xi_k; \qquad \|\nabla \cdot (c\nabla u)\|_2^2 = \sum_{k=1}^{\infty} \eta_k^2 \langle u, \xi_k \rangle_2^2.$$

In practice, we tackle the eigenvalue problem (17) via finite element methods exactly as outlined in Sect. 3.1.1 for the computation of the Dirichlet-Laplacian eigenbasis, obtaining numerical approximations of the eigenpairs. We note that, while used here as a convenient computational device to evaluate the asymptotic variances, knowledge of the SVD of the forward operator G is not assumed for the theoretical results of Sect. 2.2, nor is required for the specification of the two classes of Gaussian priors introduced in Examples 1 and 2 respectively. The theory and methodology investigated in the present article are indeed generally applicable to inverse problems where the SVD might be challenging or unfeasible to compute, or to settings where the properties of the associated eigenpairs might be unknown.

Figure 6 shows the (approximate) distributions of the plug-in posterior mean estimators $\langle \bar{f}_n, \psi \rangle_2$ for four representative test functions $\psi = \phi_2, \phi_4, \phi_8, \phi_{16}$. The plots present the histograms relative to 1000 realisations of the estimators, obtained by drawing 1000 independent collections of observations from the inverse regression model (3). For each experiment, a sample of size $n = 1000$ was drawn, with noise standard deviation $\sigma = 0.0005$. As expected from the central limit theorem (5), the distributions of the plug-in estimators $\langle \bar{f}_n, \psi \rangle_2$ exhibit a normal shape, are approximately centred around the true parameter $\langle f_0, \psi \rangle_2$, and their spread is mostly captured by the asymptotic variance $\|\nabla \cdot (c\nabla\psi)\|_2^2$.

Finally, Table 2 reports the coverage, for increasing sample sizes, of the 95%-credible intervals (6) for the same linear functionals $\langle f, \phi_j \rangle_2$, with $j = 2, 4, 8, 16$, considered in the previous set of experiments. The results were obtained by drawing 1000 independent collections, of size $n = 1000$, of observations from the inverse regression model (3), with noise standard deviation $\sigma = 0.0005$. For each random sample, a realisation of the 95%-credible intervals $\bar{\mathbf{f}}_{n,j} \pm 1.96\sqrt{\Gamma_{n,jj}}$ was obtained, and the final coverage scores were computed as the fraction of times in which the true parameters $\langle f_0, \phi_j \rangle_2$ were contained in the obtained credible intervals. As expected from the theoretical convergence result in (7), the observed coverages stabilise, as the sample size increases, around the prescribed credibility level 95%.

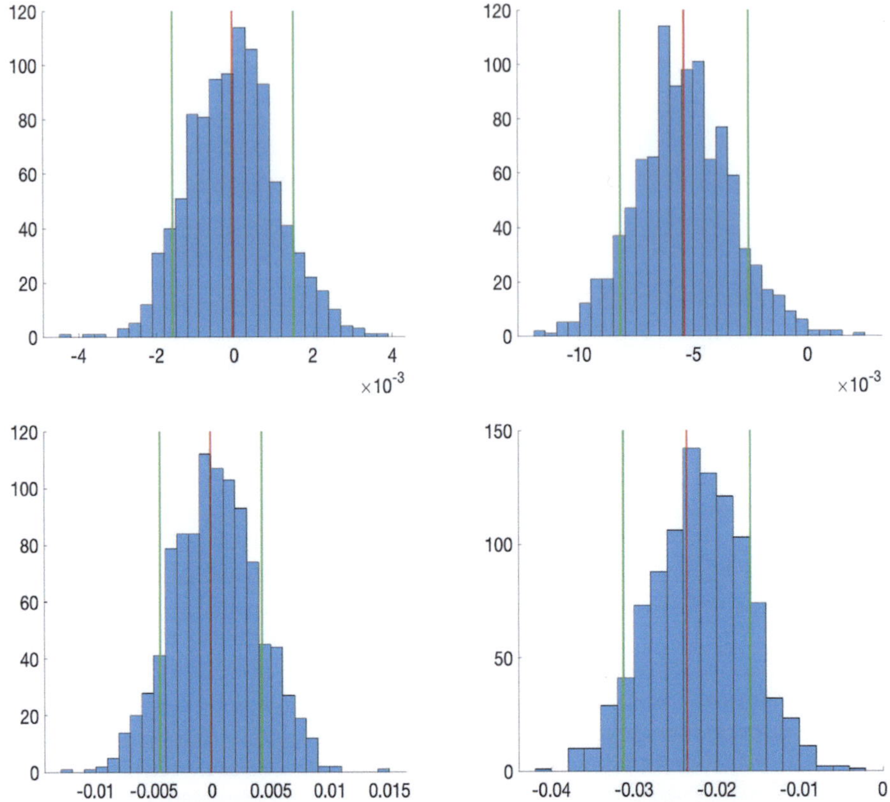

Fig. 6 Left to right, top to bottom: histograms relative to 1000 realisations of the plug-in posterior mean estimators $\langle \bar{f}_n, \phi_j \rangle_2$, for $j = 2, 4, 8, 16$. The vertical red lines identify the true parameters $\langle f_0, \phi_j \rangle_2$. The vertical green lines identify the predicted asymptotic quantiles $\langle f_0, \phi_j \rangle_2 \pm 1.96 \| \nabla \cdot (c \nabla \phi_j) \|^2 \sigma / \sqrt{n}$ (recalling the asymptotic regime $\varepsilon^2 \simeq \sigma^2 / n$)

Table 2 Observed coverages for increasing sample sizes of the 95%-credible intervals for the linear functionals $\langle f, \phi_j \rangle_2$, with $j = 2, 4, 8, 16$

n	50	100	250	500	750	1000
ϕ_2	0.885	0.921	0.954	0.969	0.947	0.962
ϕ_4	0.904	0.936	0.943	0.96	0.953	0.958
ϕ_8	0.92	0.934	0.951	0.944	0.963	0.952
ϕ_{16}	0.949	0.922	0.92	0.946	0.945	0.96

3.2 Posterior Inference with the Matérn Process Prior

Next, we consider the Matérn process priors introduced in Example 2. We discretise the parameter space by assuming that f is given by the finite sum

$$f = \sum_{m=1}^{M} f_m \varphi_m, \qquad f_1, \ldots, f_M \in \mathbb{R}, \qquad M \in \mathbb{N}, \tag{18}$$

where $\varphi_1, \ldots, \varphi_M$ are piecewise linear functions on the nodes $z_1, \ldots, z_M \in \mathcal{O}$ of a deterministic triangular mesh, uniquely identified by the property $\varphi_m(z_{m'}) = 1_{\{m=m'\}}$; see Fig. 7. Accordingly, f in (18) satisfies $f(z_m) = f_m$, and for any $x \in \mathcal{O}$ the value $f(x)$ is obtained by linearly interpolating the pairs $\{(z_m, f_m),\ m = 1, \ldots, M\}$.

Under the discretisation (18), the inverse regression model (3) can be written in matrix notation exactly as in (13), now with

$$\mathbf{G} := [G(\varphi_m)(x_i),\ i = 1, \ldots, n,\ m = 1, \ldots, M] \in \mathbb{R}^{n,M},$$

and with

$$\mathbf{f} := (f_1, \ldots, f_M)^T \in \mathbb{R}^M.$$

Thus, again, $\mathbf{Y}|\mathbf{f} \sim N_n(\mathbf{Gf}, \sigma^2\mathbf{I}_n)$. Similarly to Sect. 3.1.1, the numerical computation of the matrix $\mathbf{G}$ can be carried out with finite element methods for elliptic PDEs.

Recalling that, under the discretisation (18), $f_m = f(z_m)$ for $m = 1, \ldots, M$, and the finite dimensional distributions property (11), assigning to f a Matérn process prior with covariance $C_{\alpha,\ell}$ as in (10) corresponds to assigning to the vector $\mathbf{f}$ the M-dimensional Gaussian prior with covariance matrix

$$\mathbf{f} \sim N_M(0, \mathbf{C}), \qquad \mathbf{C} := [C_{\alpha,\ell}(z_h, z_m)]_{h,m=1}^{M} \in \mathbb{R}^{M,M}.$$

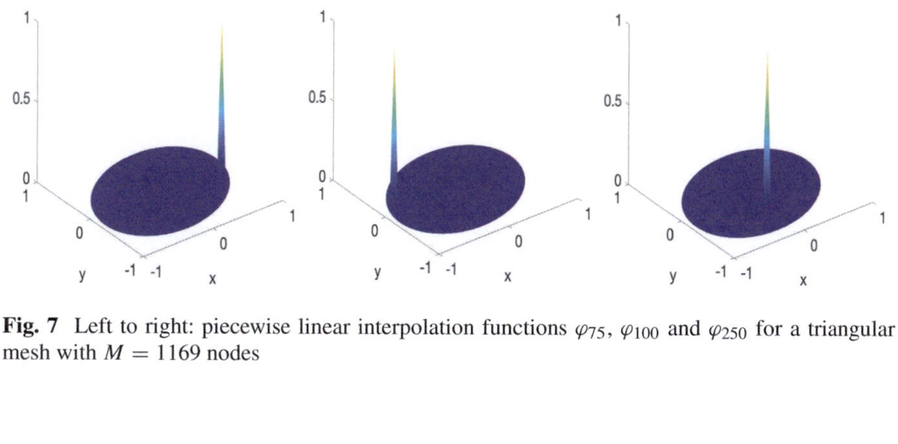

Fig. 7 Left to right: piecewise linear interpolation functions φ_{75}, φ_{100} and φ_{250} for a triangular mesh with $M = 1169$ nodes

Table 3 L^2-estimation errors achieved by the posterior mean estimator $\bar{f}_n$ arising from the Matérn process prior for increasing sample sizes

n	50	100	250	500	750	1000	2000	3000	4500
$\|\bar{f}_n - f_0\|_2$	0.30	0.30	0.18	0.12	0.13	0.10	0.086	0.076	0.067
$\|\bar{f}_n - f_0\|_2/\|f_0\|_2$	62.5%	62.5%	37.5%	25%	27.1%	20.8%	17.9%	15.6%	13.9%

The same conjugate computation as the one outlined in Sect. 3.1.1 can then be carried out, leading to the Gaussian posterior distribution $\mathbf{f}|\mathbf{Y} \sim N_M(\bar{\mathbf{f}}_n, \mathbf{C}_n)$, with posterior mean and covariance matrix respectively given by

$$\bar{\mathbf{f}}_n := \frac{1}{\sigma^2}\mathbf{C}_n\mathbf{G}^T\mathbf{Y}; \qquad \mathbf{C}_n := (\sigma^{-2}\mathbf{G}^T\mathbf{G} + \mathbf{C}^{-1})^{-1}.$$

Using the above conjugate formulae, posterior inference for the source function f based on the Matérn process prior can efficiently be implemented.

For the ground truth displayed in Fig. 1 (left), Table 3 reports the L^2-estimation errors attained by the posterior mean estimate $\bar{f}_n = \sum_{m=1}^M \bar{\mathbf{f}}_{n,m}\varphi_m$ based on an increasing number of observations from the inverse regression model (3), with noise standard deviation $\sigma = 0.0005$ (with corresponding signal-to-noise ratio $\|G(f_0)\|_2/\sigma = 37.55$). Across the experiments, the parameter space was discretised using a triangular mesh with $M = 1169$ nodes. The prior regularity parameter for the Matérn process prior was set to $\alpha = 10$, and the length-scale parameter to $\ell = 0.25$. Figure 8 shows the posterior mean estimate resulting from $n = 4500$ observations.

The results are relative to the same collection of synthetic data sets with increasing sample size employed in Sect. 3.1.2, allowing a direct comparison to the results obtained with the Gaussian series priors considered therein. Overall, the achieved L^2-estimation errors are comparable in magnitude for each sample size, albeit the performance of the Gaussian series priors was consistently slightly better. It is plausible that such small discrepancy is caused by finite sample effects, prior tuning and the various numerical approximations.

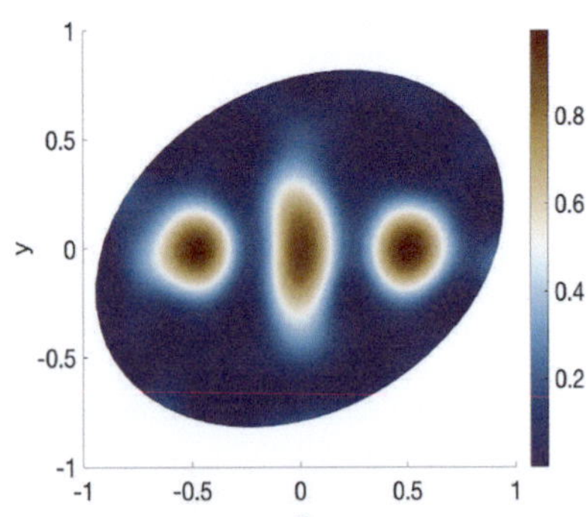

Fig. 8 The posterior mean estimate of the source function f based on a Matérn process prior with parameters $\alpha = 10$, $\ell = 0.25$

3.3 *Further Numerical Experiments*

3.3.1 Sensitivity to the Noise Variance

For the empirical results presented in Sects. 3.1 and 3.2, a fixed noise standard deviation $\sigma = 0.0005$ in the inverse regression model (3) was used (with corresponding signal-to-noise ratio $\|G(f_0)\|_2/\sigma = 37.55$). Here, we provide a brief investigation of the sensitivity of the considered methodology to the value of σ, performing a set of experiments with decreasing standard deviation from $\sigma = 0.01$ (for which $\|G(f_0)\|_2/\sigma = 1.8774$) to $\sigma = 0.0001$ (for which $\|G(f_0)\|_2/\sigma = 187.74$), based on the same domain and ground truth used previously. Across the experiments, the sample size was kept fixed at $n = 4500$.

For concreteness, we focus on the Gaussian series priors from Sect. 3.1, with the same prior tuning employed therein; similar results may be obtained with the Matérn process priors. The L^2-estimation error associated to the resulting posterior mean estimates are shown in Table 4. Unsurprisingly, these were observed to decrease monotonically as the signal-to-noise ratio increased. In particular, at the lowest value $\sigma = 0.0001$, the L^2-estimation error may be seen to approach the L^2-approximation error resulting from projecting f_0 onto the employed basis, which is equal to 0.0486.

3.3.2 Inference with Unknown Noise Variance

We conclude the simulation study considering the important practical scenario where the noise standard deviation σ is itself unknown and needs to be estimated from the data. Given observations $\{(Y_i, X_i)\}_{i=1}^n$ from the inverse regression model (3), we undertake the simple 'empirical Bayes' approach of obtaining a preliminary estimate $\hat{\sigma}_n$ of σ, and then carry over the methodology laid out in Sects. 3.1 and 3.2 with σ replaced by $\hat{\sigma}_n$. Alternatively, we note that a joint Bayesian model for f and σ in (3) could be considered by endowing σ with a prior distribution. For example, an independent inverse-gamma distribution would lead (conditionally given f) to a conjugate posterior distribution, whereupon joint posterior samples for the pair (f, σ) could readily be obtained via a Gibbs sampler, alternating draws from the Gaussian posterior distribution of $f|\{(Y_i, X_i)\}_{i=1}^n, \sigma$ and draws from the inverse-

Table 4 L^2-estimation errors achieved by the posterior mean estimator $\bar{f}_n$ arising from the Gaussian series prior for decreasing noise standard deviation

σ	0.01	0.005	0.0025	0.001	0.0005	0.0001
$\|G(f_0)\|_2/\sigma$	1.88	3.75	7.51	18.77	37.55	187.74
$\|\bar{f}_n - f_0\|_2$	0.21	0.16	0.14	0.078	0.060	0.049
$\|\bar{f}_n - f_0\|_2/\|f_0\|_2$	43.75%	33.33%	29.17%	16.25%	12.5%	10.20%

Table 5 Inferential results for the difference-based estimator $\hat{\sigma}_n$ of the noise standard deviation and the the empirical Bayes posterior mean estimator $\hat{f}_n$, for increasing sample sizes

n	1000	2000	3000	4500
$\hat{\sigma}$	0.0034	0.0024	0.0023	0.00072
$\|\hat{f}_n - f_0\|_2$	0.18	0.14	0.12	0.063
$\|\hat{f}_n - f_0\|_2/\|f_0\|_2$	37.5%	29.17%	25%	13.12%

gamma posterior distribution of $\sigma|\{(Y_i, X_i)\}_{I=1}^n, f$. For brevity, we will not pursue this approach further here.

Several strategies have been proposed in the literature for variance estimation in nonparametric regression models, ranging from residual-based estimators using kernel smoothing [22] and splines [46], to difference-based estimators [38]. See [5] for an overview. Here, we will consider the difference-based method proposed in [38], estimating σ in model (3) by

$$\hat{\sigma}_n := \sqrt{\hat{\sigma}_n^2}, \qquad \hat{\sigma}_n^2 := \frac{1}{2(n-1)} \sum_{i=2}^{n} (Y_i - Y_{i-1})^2.$$

Based on $\hat{\sigma}_n$, the 'empirical Bayes posterior mean' estimate $\hat{f}_n$ arising from a Gaussian series prior or a Matérn process prior can then be readily computed exactly as described in Sects. 3.1 and 3.2 respectively, replacing σ with $\hat{\sigma}_n$ in the relevant conjugate formulae.

Table 5 summarises the inferential results obtained with the difference-based estimation procedure for increasing sample sizes. For these experiments, the noise standard deviation was set to $\sigma = 0.0005$, and a Gaussian series priors with the same tuning as in Sects. 3.1.2 and 3.3.1 was used. The results show a progressive improvement in the reconstruction quality for both the noise standard deviation σ and the unknown source function f. In particular, for the largest considered sample size $n = 4500$, the L^2-estimation error $\|\hat{f}_n - f_0\|_2$ resulted to be only marginally higher than the one obtained under the same experimental conditions (and prior tuning) in the context of the empirical results presented in Sect. 3.1.2 (cf. Table 1), for which knowledge of the value of σ was assumed.

4 Summary and Discussion

In this article we have considered the nonparametric Bayesian approach with Gaussian priors to linear inverse problems, focusing on the important example of source identification in elliptic PDEs. The main advantages of the considered methodology lie in its modelling flexibility, its ease of implementation (cf. the conjugate formulae (15) and (16)), as well as its theoretical guarantees on estimation and uncertainty quantification (cf. Sect. 2.2). The performance of the approach has been investigated in a numerical simulation study (cf. Sect. 3) under two distinct prior models (Gaussian series and Matérn process priors), both for which excellent reconstruction results have been obtained.

The present work also raises various related research questions. Firstly, it is of interest and practical importance to further explore the setting where the noise standard deviation σ in the inverse regression model (3) is unknown. While the simple difference-based estimator considered in Sect. 3.3.2 has proved effective, several competing approaches, including the joint conjugate Bayesian model outlined in Sect. 3.3.2, could be investigated. Furthermore, a related interesting question concerns the extensions of the theoretical results presented in Sect. 2.2 to the setting with unknown variance; see e.g. [26] for related results in a direct regression model.

Lastly, let us mention the important issue of specifying the hyperparameter values for the considered prior distributions, namely the truncation level and regularity in the Gaussian series (14), and the smoothness and length-scale parameters in the Matérn covariance kernel (10). There is by now a vast literature investigating the methodological and theoretical aspects of empirical and hierarchical Bayesian approaches to fully data-driven selection of the hyperparameters; see [3, 28, 39, 45] and the many reference therein. Investigating the implications and performance of these methods in the context of the observation model and prior distributions considered in the present article is an interesting problem for future research.

Acknowledgements The Author is grateful to three anonymous referee for many helpful comments that lead to an improvement of the article.

Competing Interests This research has been partially supported by MUR, PRIN project 2022 CLTYP4. The Author also gratefully acknowledges the support of "de Castro" Statistics Initiative, Collegio Carlo Alberto, Torino. There are no conflicts of interest to declare that are relevant to the content of this chapter.

References

1. Abraham, K., Nickl, R.: On statistical Calderón problems. Math. Stat. Learn. **2**(2), 165–216 (2019)
2. Adavani, S.S., Biros, G.: Fast algorithms for source identification problems with elliptic PDE constraints. SIAM J. Imag. Sci. **3**(4), 791–808 (2010)
3. Agapiou, S., Bardsley, J.M., Papaspiliopoulos, O., Stuart, A.M.: Analysis of the Gibbs sampler for hierarchical inverse problems. SIAM/ASA J. Uncertain. Quant. **2**(1), 511–544 (2014)
4. Agapiou, S., Stuart, A.M., Zhang, Y.-X.: Bayesian posterior contraction rates for linear severely ill-posed inverse problems. J. Inverse Ill-Posed Probl. **22**(3), 297–321 (2014)
5. Alharbi, Y.F.M.: Error variance estimation in nonparametric regression models. Ph.D. thesis, University of Birmingham (2013)
6. Arridge, S., Maass, P., Öktem, O., Schönlieb, C.-B.: Solving inverse problems using data-driven models. Acta Numer. **28**, 1–174 (2019)
7. Baumeister, J.: Inverse problems in finance. In: Recent Developments in Computational Finance: Foundations, Algorithms and Applications. World Scientific, pp. 81–157 (2013)

8. Benning, M., Burger, M.: Modern regularization methods for inverse problems. Acta Numerica **27**, 1–111 (2018)
9. Bertero, M., Piana, M.: Inverse problems in biomedical imaging: modeling and methods of solution, pp. 1–33. In: Complex Systems in Biomedicine. Springer Italia, Milan (2006)
10. Brown, L.D., Low, M.G.: Asymptotic equivalence of nonparametric regression and white noise. Ann. Statist. **24**(6), 2384–2398 (1996)
11. Castillo, I., Nickl, R.: Nonparametric Bernstein-von Mises theorems in Gaussian white noise. Ann. Statist. **41**(4), 1999–2028 (2013)
12. Collins, M., Kuperman, W.: Inverse problems in ocean acoustics. Inverse Probl. **10**(5), 1023 (1994)
13. Elvetun, O.L., Nielsen, B.F.: A regularization operator for source identification for elliptic PDEs. Inverse Probl. Imag. **15**, 4 (2021)
14. Evans, L.C.: Partial differential equations. In: Graduate Studies in Mathematics, vol. 19, 2nd edn. American Mathematical Society, Providence, RI (2010)
15. Ghosal, S., van der Vaart, A.W.: Fundamentals of Nonparametric Bayesian Inference. Cambridge University Press, New York (2017)
16. Giné, E., Nickl, R.: Mathematical Foundations of Infinite-Dimensional Statistical Models. Cambridge University Press, New York (2016)
17. Giordano, M.: Bayesian nonparametric inference in PDE models: asymptotic theory and implementation. JSM Proc. Zenodo **2023**, 1–17 (2023)
18. Giordano, M., Kekkonen, H.: Bernstein-von Mises theorems and uncertainty quantification for linear inverse problems. SIAM/ASA J. Uncertain. Quantif. **8**(1), 342–373 (2020)
19. Giordano, M., Nickl, R.: Consistency of Bayesian inference with Gaussian process priors in an elliptic inverse problem. Inverse Probl. **36**(8), 085001–085036 (2020)
20. Giordano, M., Ray, K.: Nonparametric Bayesian inference for reversible multidimensional diffusions. Ann. Stat. **50**(5), 2872–2898 (2022)
21. Gugushvili, S., van der Vaart, A., Yan, D.: Bayesian linear inverse problems in regularity scales. Ann. de l'Institut Henri Poincaré, Probabilités et Statistiques **56**(3), 2081–2107 (2020)
22. Hall, P., Marron, J.: On variance estimation in nonparametric regression. Biometrika **77**(2), 415–419 (1990)
23. Haroske, D.D., Triebel, H.: Distributions, Sobolev Spaces. EMS Press, Elliptic Equations (2007)
24. Isakov, V.: Inverse problems for partial differential equations. In: Applied Mathematical Sciences, vol. 127, 3rd edn. Springer, Cham (2017)
25. Kaipio, J., Somersalo, E.: Statistical and computational inverse problems. In: Applied Mathematical Sciences. Springer-Verlag New York (2004)
26. Kejzlar, V., Son, M., Bhattacharya, S., Maiti, T.: A fast and calibrated computer model emulator: an empirical Bayes approach. Stat. Comput. **31**(4), 49 (2021)
27. Kekkonen, H., Lassas, M., Siltanen, S.: Posterior consistency and convergence rates for Bayesian inversion with hypoelliptic operators. Inverse Probl. **32**(8), 085005 (2016)
28. Knapik, B., Szabò, B., van der Vaart, A.W., van Zanten, H.: Bayes procedures for adaptive inference in inverse problems for the white noise model. Probab. Theory Relat. Fields **164**, 771–813 (2015)
29. Knapik, B., van der Vaart, A.W., van Zanten, J.H.: Bayesian inverse problems with Gaussian priors. Ann. Statist. **39**(5), 2626–2657 (2011)
30. Knapik, B.T., van der Vaart, A.W., van Zanten, J.H.: Bayesian recovery of the initial condition for the heat equation. Comm. Statist. Theory Methods **42**(7), 1294–1313 (2013)
31. Lehtinen, M.S.: On statistical inversion theory. Theory Appl. inverse probl. **67**, 46–57 (1988)
32. Lehtinen, M.S., Paivarinta, L., Somersalo, E.: Linear inverse problems for generalised random variables. Inverse Probl. **5**(4), 599 (1989)
33. Lions, J.L., Magenes, E.: Non-Homogeneous Boundary Value Problems and Applications, 1st edn. Grundlehren der mathematischen Wissenschaften. Springer-Verlag, Berlin Heidelberg (1972)

34. Monard, F., Nickl, R., Paternain, G.P.: Efficient nonparametric Bayesian inference for X-ray transforms. Ann. Statist. **47**(2), 1113–1147 (2019)
35. Monard, F., Nickl, R., Paternain, G.P.: Consistent inversion of noisy non-Abelian X-ray transforms. Comm. Pure Appl. Math. **74**(5), 1045–1099 (2021)
36. Nickl, R., van de Geer, S., Wang, S.: Convergence rates for penalized least squares estimators in PDE constrained regression problems. SIAM/ASA J. Uncertain. Quantif. **8**(1), 374–413 (2020)
37. Rasmussen, C.E., Williams, C.K.I.: Gaussian processes for machine learning. In: Adaptive Computation and Machine Learning. MIT Press, Cambridge, MA (2006)
38. Rice, J.: Bandwidth choice for nonparametric regression. Ann. Stat. 1215–1230 (1984)
39. Rousseau, J., Szabo, B.: Asymptotic Behaviour of the empirical Bayes posteriors associated to maximum marginal likelihood estimator. Ann. Stat. **45**(2), 833–865 (2017)
40. Snieder, R., Trampert, J.: Inverse problems in geophysics. In: Wirgin, A. (Ed.) Wavefield Inversion, pp. 119–190. Springer Vienna (1999)
41. Stuart, A.M.: Inverse problems: a Bayesian perspective. Acta Numer. **19**, 451–559 (2010)
42. Tarantola, A.: Inverse Problem Theory and Methods for Model Parameter Estimation. Society for Industrial and Applied Mathematics (SIAM), Philadelphia, PA (2005)
43. Tarantola, A., Valette, B., et al.: Inverse problems = quest for information. J. Geophys. **50**(1), 159–170 (1982)
44. Taylor, M.E.: Partial Differential Equations I. Springer, New York, NY (2011)
45. Teckentrup, A.L.: Convergence of gaussian process regression with estimated hyper-parameters and applications in Bayesian inverse problems. SIAM/ASA J. Uncertain. Quantif. **8**(4), 1310–1337 (2020)
46. Wahba, G.: Improper priors, spline smoothing and the problem of guarding against model errors in regression. J. Roy. Stat. Soc. Ser. B: Stat. Methodol. **40**(3), 364–372 (1978)
47. Yeh, W.W.-G.: Review of parameter identification procedures in groundwater hydrology: the inverse problem. Water Resour. Res. **22**(2), 95–108 (1986)

Phase I Dose Escalation Trials in Cancer Immunotherapy: Modifying the Bayesian Logistic Regression Model for Cytokine Release Syndrome

Matt Chapman-Rounds and Miguel Pereira

Abstract We extend Bayesian Logistic Regression to model the dose-toxicity relationship in the setting of phase I dose-escalation/dose-finding trials for cancer immunotherapies. Immunotherapy drugs are associated with Cytokine Release Syndrome, a systemic immune reaction that can be mitigated when initial lower doses of the drug are administered to generate immune tolerance. This complicates the standard dose-finding problem; we now need to find both the optimal safe dose and the dose regimen that allows patients to quickly and safely reach that dose without CRS. As part of solving this methodological challenge, we show how to (i) jointly model CRS and non-CRS toxicities, and (ii) generalise the BLRM to model dose regimens, in addition to single doses.

1 Introduction

The goals of a phase I, first-in-human, dose-finding clinical trial are to collect information on the safety and tolerability of a new drug, identify the maximum tolerated dose (MTD), and identify one or more doses to assess in a phase II study (recommend phase II doses or RP2D) [5].

The MTD is the highest dose which can be administered to humans whilst balancing efficacy with toxicity [5, 9]. Toxicity is defined as the probability that a subject experiences a serious adverse event, also called a dose-limiting toxicity or DLT. Generally, both drug efficacy and drug toxicity increase with dose. In order to find the MTD, the standard approach of a phase I trial is to start at a low dose and carefully escalate to higher doses in subsequent groups of participants (usually size 3–6). As the dose increases, DLTs will start to occur, and at some point the trial will stop increasing the dose when it is concluded that the toxicity of any further increase will be too high. This decision is often supported by an explicit model of the dose-toxicity relationship, such as the Bayesian Logisitic Regression Model (BLRM) [14].

M. Chapman-Rounds (✉) · M. Pereira
Cogitars UK Ltd, Cambridge, UK
e-mail: matt.chapman-rounds@cogitars.com

A. Avalos-Pacheco et al. (eds.), *New Trends in Bayesian Statistics*,
Springer Proceedings in Mathematics & Statistics 511,
https://doi.org/10.1007/978-3-031-99009-0_3

The advent of novel treatments in oncology has lead to the development of multiple immuno-therapeutic agents whose action is to stimulate the immune system to act against the tumour cells. An import toxicity caused by these drugs is Cytokine Release Syndrome (CRS), a systemic inflammatory reaction that occurs when many immune cells are activated and release large amount of cytokines into the body [4].

CRS is more likely to occur at lower doses in naïve patients versus patients whose immune system has already been primed to the drug. To avoid CRS therefore, subjects are first exposed to lower doses of a drug compound in preparation to receiving the target dose. As such, modelling dose-toxicity in this setting poses three novel problems:

1. Instead of modelling a single dose level, we need to model a dose *regimen* whereby patients receive several lower doses of the drug before reaching the target dose. The goal is now to find the maximum tolerated dose regimen (MTD-R), which is a much more challenging problem than finding the MTD because we need to consider multiple additional factors including the number of prior doses, and the shape of the dosing curve.
2. To capture the ameliorating effects of exposing subjects to preparatory lower doses, we need to model both CRS toxicities and non-CRS toxicities, and then combine these models to estimate the joint posterior probability of toxicity.
3. In case of toxicity, we need to identify the step(s) in the dose regimen that need(s) adjustment, to better drive dose regimen-escalation decisions during the phase I trial.

This work focusses on generalising the BLRM to accommodate modelling of both CRS and non-CRS toxicities in phase I trials for immunotherapy drugs, and secondly on generalising the BLRM to the multivariate setting of dose-*regimen* finding, from the standard univariate dose finding setting. Our approach builds on an adaptation of the methodology proposed by Gerard et al. [2].

2 Related Work

Historically, the most widely used method for dose escalation has been the 3+3 design. In this design, cohorts of 3 patients are treated at increasing dose levels with an option to expand to 6 patients at a given cohort. It has simple and clearly defined rules, thus facilitating its widespread application [5]. However, compared to other methods, the 3+3 design has poor operating characteristics showing a higher probability of allocating patients to sub-therapeutic dose levels and a lower accuracy in defining the MTD [9].

Alternative designs include rule-based designs like the mTPI [12], the keyboard design (mTPI2) [8, 13], and the Bayesian optimal interval (BOIN) [10]; and model-based designs such as Bayesian response adaptive models [7].

The mTPI and the keyboard desgin are interval-based designs that define clear boundaries that lead to a decision of 'Escalation', 'De-escalation' or 'Stay' according to the number of DLTs observed in a cohort. For both methods, it is possible to create decision tables given the size of the cohort and the number of DLT events, which is very useful to streamline dose escalation. The difference between these two methods is in how the intervals are defined, where the mTPI2 addresses some of the pitfalls of the mTPI design (itself an improvement on the TPI design) that led to suboptimal rules [13].

The BOIN is a generalisation of the 3+3 and accelerated titration designs which demonstrates superior operating characteristics [11]. It contains clear escalation rules and has flexibility in the number of cohorts to be included and in the targeting rate [10]. In addition, there are multiple extensions of the BOIN to other cases such as time-to-DLT, modelling different toxicity grades, or concomitant modelling of safety and efficacy [11].

Bayesian response adaptive models for phase 1 studies have been advocated for by the European Medicines Agency adopted guideline on small populations [3], by Rogatko and colleagues [1, 7], and were one of the key elements of the FDA's Critical Path Initiative.

Of these, the BLRM is an increasingly common methodology used to support dose escalation [14]. The BLRM with Escalation with Overdose Control (EWOC) performs significantly better than the 3+3 design, showing better targeting rates (~30% in the 3+3 vs. ~60% in the BLRM) with fewer patients treated above the MTD [1]. Additionally, because the BLRM considers previously observed data it provides added flexibility as it allows cohorts of different sizes along with escalation to intermediate dose levels [14].

3 Models

The standard BLRM defines the probability of DLT, $P(DLT)$ at dose D as:

$$P(DLT) = \text{logit}^{-1}(\log(\alpha) + \beta * \log(D/D^*)), \tag{1}$$

where α and β are the parameters of the model and D^* is the reference dose such that the odds of DLT at dose D^* are equal to α.

At each stage in a standard phase I trial, the posterior distributions of parameters α and β would be updated via Bayes' Rule based on the incidence of DLTs observed in subjects thus far. These posteriors would then be used to generate the probability of $P(DLT)$, i.e. the posterior probability density over possible true DLT rates at the next dose, or doses, under consideration. Doses can then be described as being in the underdose, overdose, or target dose range. Using thresholds often used in Oncology trials, we define these ranges as follows:

- **Overdose:** Probability $P(DLT) >= 0.33$ is greater than 0.25.
- **Target dose:** Probability $0.16 <= P(DLT) < 0.33$ is greater than 0.5, and probability $P(DLT) >= 0.33$ is less than 0.25.

- **Underdose:** Probability $P(DLT) < 0.16$ is greater than 0.5, and probability $P(DLT) >= 0.33$ is less than 0.25.

If the current dose satisfies the requirements for target dose, and the next possible dose is an overdose, then escalation stops and the dose is declared the MTD. If the next dose is an overdose, but the current dose does not satisfy the condition to be the target dose, the dose might still be considered as the RP2D. Alternatively, escalation to a slightly higher, intermediate dose level may be tested to accurately ascertain the MTD.

3.1 Modified BLRM

When CRS is a concern, our goal is to find the maximum tolerable dose *regimen* (MTD-R), rather than modelling simply the relationship between dose and toxicity. To do this, we need to model the relationship between a succession of doses and the predicted toxicity.

Our approach is to track two distinct types of adverse events—those due to CRS, and those which are due to other types of toxicity (non-CRS adverse events). The probabilities of each type of event are modelled separately, and assumed to be independent.[1] Non-CRS adverse events are modelled using a standard BLRM, whilst CRS adverse events are modelled by considering the probability of a CRS adverse event at each stage of the dose regimen under consideration.

Consider a dose regimen which includes an escalation period and a steady state period. For example, a four-step regimen might include the doses 5, 10, 25, 50, 50, 50 at intervals of several days to complete a cycle whose 'steady state' or target dose dose is 50 μg. We call the period between two doses a 'stage'; for example, the above is a four-step dose regimen (four distinct doses), but has 6 stages (the final stage will extend from the final dose until end of the safety follow-up).

As with the standard approach, the goal is to generate the posterior probability of $P(DLT)$ for any dose regimen, and determine if this regimen corresponds to a target dose, underdose, or overdose. Figure 1 plots example distributions after 0 out of 3 subjects experienced a DLT when given the first regimen, and 1 out of 4 subjects experienced a DLT when given the second dosing regimen. Regimens are indexed by their steady state dose; the third and next regimen to be considered in Fig. 1 is the four-step regimen described above, with a steady state dose of 50 μg.

[1] In practice, this is not quite true: for example, a subject experiencing a CRS DLT will withdraw from the study, reducing the chance they experience a non-CRS DLT, and visa versa. For the purposes of demonstrating our basic model we have elected to forgo modelling any dependence relationship between the two types of events: this would be the subject of further work.

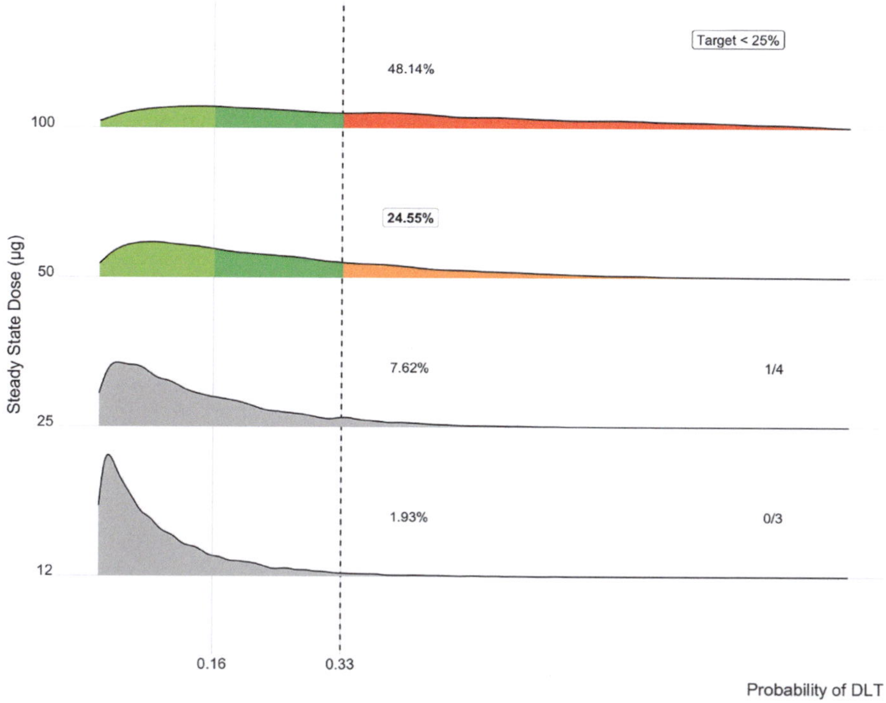

Fig. 1 Posterior densities after 1 DLT observed out of 4 subjects in the second regimen, and 0 DLTs observed out of 3 subjects in the first regimen. Red colour indicates that the probability mass $>= 0.33$ is > 0.25

3.1.1 CRS Model

We assume that for each stage i, the principle positive correlate with probability of CRS in stage i is $C_{max,i}$, the maximum concentration reached after dose administration. Whilst we lack the pharmacokinetic data to capture this directly (this data is seldom available at the time of dose escalation), we will take the approach of approximating it using the dose at the start of stage i, D_i, i.e.

$$C_{max,i} \approx D_i \tag{2}$$

This is counter-balanced by the exposure to the drug prior to stage i (since exposure decreases the likelihood of CRS). We approximate this exposure using the 'area under the dose', AUD_i,

$$\text{AUD}_i = \sum_{k=1}^{k=(i-1)} T_k * D_k \tag{3}$$

where T_k is the length of stage k.

For any stage i we define the probability that a CRS-related DLT occurs within that stage, $P(\text{CRS}_i)$ as

$$P(\text{CRS}_i) = \text{logit}^{-1}\Big(\log(\alpha_{1i}) + \gamma_{1i} * \log(D_i/D_i^*) - \beta_{1i} * \log(\text{AUD}_i/\text{AUD}_i^*)\Big) \tag{4}$$

where AUD_i^* is the AUD for the ith stage of a selected reference regimen, and D_i^* is the dose for the ith stage of that reference regimen. Note that it is possible to choose references from different dose regimens—however, for simplicity, we will assume here that a single dose regimen is chosen as reference, and the same reference is used across all stages. If this assumption is not kept, then α no longer has a particularly informative interpretation.

The probability that no CRS occurs in a subject is the joint probability of CRS not occuring in each stage which, assuming S stages, we can write as $P(\text{no CRS}_1, ..., \text{no CRS}_i, ..., \text{no CRS}_S)$.

Firstly, we note that $P(\text{no CRS}_i)$ is independent of all $P(\text{no CRS}_j)$, $j > i$. This follows from the direction of causality.

Secondly, we note that $P(\text{no CRS}_i) \perp\!\!\!\perp P(\text{no CRS}_{(i-1)}|\text{AUD}_i)$, i.e. that the probability of CRS/no CRS for stage i is conditionally independent of the probability of CRS/no CRS for the previous stage $(i-1)$ given the dose exposure up until the end of that previous stage. This follows from the fact that our model's calculation of AUD_i includes all components contributing to the probability of CRS at the previous stage ($D_{(i-1)}$ and $\text{AUD}_{(i-1)}$). Given this Markov property for dose stages, it follows that $P(\text{no CRS}_i) \perp\!\!\!\perp P(\text{no CRS}_j|\text{AUD}_i)$, for all $j < i$.

Hence we can write

$$P(\text{no CRS}_1, ..., \text{no CRS}_i, ..., \text{no CRS}_S) = \prod_{i=1}^{i=S} P(\text{no CRS}_i) \tag{5}$$

$$= \prod_{i=1}^{i=S} (1 - P(\text{CRS}_i)) \tag{6}$$

which is the probability that no CRS occurs. As CRS either occurs or does not occur, the probability that CRS occurs is simply

$$P(\text{CRS}) = 1 - \prod_{i=1}^{i=S} (1 - P(\text{CRS}_i)), \tag{7}$$

which completes our model of CRS related DLTs.

3.1.2 Non-CRS Model

This model is similar to the standard BLRM, with the exception that it uses the cumulative dose over the regimen, $D_{cum} = \sum_{i=1}^{i=S} D_i$, to predict the probability that any other type of DLT occurs. Note that this model does not consider where DLTs fall in terms of stages, but rather estimates the probability of a non-CRS DLT, $P(\text{non-CRS})$, occuring at any time. We write

$$P(\text{non-CRS}) = \text{logit}^{-1}(\log(\alpha_2) + \delta_2 * \log(D_{cum}/D^*_{cum})), \tag{8}$$

where D^*_{cum} is the cumulative dose over the reference regimen, which we have taken to be the same reference regimen as that used in the CRS model, above. This completes our model of non-CRS-related DLTs.

3.1.3 Joint Model

The probability of any DLT occuring is the probability that either a CRS or non-CRS DLT occurs. This is 1 minus the probability that neither a CRS or non-CRS DLT occurs. We have assumed that CRS and non-CRS DLTs, as modelled, are independent. Hence

$$\begin{aligned} P(\text{DLT}) &= 1 - P(\text{no CRS}) * P(\text{no non-CRS}) \\ &= 1 - \Big(\prod_{i=1}^{i=S}(1 - P(\text{CRS}_i)) * \big(1 - \text{logit}^{-1}(\log(\alpha_2) + \delta_2 * \log(D_{cum}/D^*_{cum}))\big)\end{aligned} \tag{9}$$

This completes our joint model of DLTs.

3.2 *Prior Elicitation*

When performing Bayesian analysis, it is necessary to specify priors. As a phase I trial generally marks the first introduction of the study drug to humans, there is little information about the dose-tolerance relationship for CRS DLTs (characterised by parameters $[\alpha_{1i}]$, $[\beta_{1i}]$, $[\gamma_{1i}]$, $i \in \{1, ..., S\}$) and non-CRS DLTs (parameters α_2, δ_2). The uncertainty about this relationship is therefore specified as a set of minimally informative priors.

The priors are defined such that for the prospective dose regimens defined at the start of the trial we expect to have minimal toxicity at the lowest dose regimen, and some toxicity (e.g. $P(DLT) = 25\%$) at the highest regimen.

The prospective dose regimens are selected by pharmacokinetic and clinical experts based on available data from animal studies and other sources of information. The intent is to specify a range of regimens such that the first dose regimen will be safe, but by the highest regimen we should have encountered some DLTs. If no DLTs have been encountered by the end of the study, but efficacy is still increasing with dose, then this indicates that too many subjects have been given low-efficacy doses, and the escalation should have been more rapid.

Setting our priors such that we expect some toxicity in the higher regimens reflects the belief that the dose regimens have been correctly chosen such that we will escalate from a safe dose to a dose which begins to produce DLTs, and hence that the prospective range contains the MTD-R.

3.2.1 Standard BLRM

In the context of the standard BLRM, for a range of prospective doses from D_1 to D_N, we might assume that $P(DLT|D_1) = 0.01$ and $P(DLT|D_N) = 0.25$. The equation for $P(DLT)$ for the standard BLRM only has two parameters α and β (see Eq. 1); assuming α and β are Gaussian distributed we can use these two assumed toxicities to solve for point estimates for prior means μ_α and μ_β. Setting the vector of standard deviations $[\sigma_\alpha, \sigma_\beta]$ to reasonably large values, e.g. [2, 1] gives us a first estimate of the prior.

Following this step, the behaviour of the BLRM under this initial prior estimate is checked by simulation. Several basic scenarios which might be examined include:

- If no DLTs have been observed to date, the BLRM should always recommend to escalate (i.e. the next dose under consideration should be estimated to be either underdose or target dose by the BLRM).
- If 3/3 DLTs are observed at any dose, the BLRM should recommend to de-escalate the dose (i.e. the current dose and the next dose under consideration should be estimated to be overdoses), possibly by more than one dose level.
- If 1/3 DLTs or 2/3 DLTs are observed at any dose, the BLRM should recommend either expansion or de-escalation. Which recommendation is appropriate would depend on several factors such as the relative size of the jump from the previous dose, and would be subject to the judgment of the expert eliciting the prior.

If the required behaviour is not met, then the prior can be adjusted by changing the assumed true probability of DLT at D_N, and re-solving.

3.2.2 Extended BLRM

Rather than two parameters, the extended BLRM has $(3 * S) + 2$ parameters. As such, assuming low toxicity at the starting dose regimen and high toxicity at a final or high dose regimen will not allow us to estimate the means of Gaussian priors for each parameter (as the matrix of the two equations generated by this assumption will be non-invertable).

To handle this, we first treat the probability of CRS and non-CRS DLTs separately. This has the advantage of allowing us some flexibility in the assumed toxicities (for example, we might expect the probability of DLTs due to non-CRS related causes to be higher than that of those due to CRS). As non-CRS DLTs are modelled by only two parameters, we can derive an initial estimate of the prior for the non-CRS DLT BLRM in the same way as for the standard BLRM.

To gain an initial estimate of the prior for the model of CRS DLTs, our approach is to assume that the CRS model has the form $\log(\alpha_{1i}) + \delta_{1i}[\log(D_i/D_i^*) + \log(\text{AUD}_i/\text{AUD}_i^*)]$. We assume that toxicity in the final dose regimen decreases linearly with stage (as inceased exposure should reduce the probability of CRS), and solve the two-point problem for α and δ for each stage. We then set $\gamma_{1i} = \delta_{1i} + \epsilon M_D$, and $\beta_{1i} = \delta_{1i} - \epsilon M_{aud}$, where ϵ is some constant (here 0.01), and M_{**} is a scaling factor which adjusts ϵ roughly by the relative magnitude of mean AUD to mean D across all regimens. Again, we assume Gaussian priors with these point estimates as the means, and wide standard deviations. Once the initial prior estimates have been established, the priors are adjusted until the model exhibits satisfactory predicted behaviour as with the standard BLRM.

This approach works well in practice, but is currently an open area of research. Finding a prior which is well behaved is something of an artform: the above method to find point estimates of the parameter prior is extremely heuristic, and relies on the fact that the priors are kept minimally informative by the addition of noise for its validity. Mixture priors may often be used if there are competing sources of information, but this adds a further layer of complexity beyond the scope of this work.

4 Methods

To demonstrate the improved efficacy of the modified BLRM in the setting where dose regimens are being used in place of single doses, compared with the standard BRLM (as a leading example of current alternative models), we simulate the operating characteristics of the models across three scenarios:

1. **Priors Correct**: In this scenario the priors of the BLRM are correct. The number of DLTs in a simulated trial are generated by drawing a sample from the prior. This simulates a scenario where we have elicited priors which reflect the true dose-tolerance relationship.
2. **High Toxicity**: The priors of the BLRM are an under-estimate. The number of DLTs in a simulated trial are generated with probability 25% more than predicted by the prior. This is a scenario where we have elicited priors which consistently under-estimate the probability of toxicity; the BLRM should swiftly discard its priors in favour of evidence of high toxicity.
3. **Low-High Toxicity**: The priors of the BLRM are used to generate simulated trial data for the first four regimens under consideration. However, in the last three regimens the toxicity spikes sharply: the true probability of DLT for these regimens is 25% higher than that predicted by the prior. This simulates a worst case scenario where both prior and evidence from lower dose regimens support

Table 1 Prospective Dose Regimens for simulation. 'DR' = Dose. All values in μg

	Day 1	Day 5	Day 10	Day 14	Day 21
DR1	1	2	5	15	15
DR2	3	6	20	75	75
DR3	10	20	50	250	250
DR4	25	50	150	550	550
DR5	25	50	300	1000	1000
DR6	25	50	300	1500	1500
DR7	25	50	300	2850	2850

a particular dose-tolerance relationship, but this does not generalise correctly to higher dose regimens. The BLRM should quickly adjust its posteriors midway through the simulated trial to reflect the new evidence at higher doses.

Each scenario assumes that we have pre-selected 7 regimens with increasing dose. Each regimen has four distinct doses ('steps') across five stages. Table 1 outlines the prospective dose regimens used for simulation. For each scenario, 500 trials were simulated as described in Algorithm 1, where DR_C denotes the current regimen under consideration.

Algorithm 1 Simulation algorithm for a single trial

1. Set $DR_C = DR_1$.
2. Sample CRS and non-CRS DLTs separately from the true scenario distribution, for 3 subjects, at DR_C.
3. Using sample data, update BLRM via Bayes' rule to obtain new posterior, and use this posterior to determined which of the prospective regimens are **target dose**, **underdose**, or **overdose**.
4. Decision step:

if DR_{C+1} is not an **overdose then**
    set $DR_C = DR_{C+1}$,
    go to 2.
else if DR_{C+1} is an **overdose then**
    if DR_C is **target dose then**
        end trial with MTD = DR_C.
    else if DR_C is **underdose then**
        go to 2.
    else if DR_C is **overdose then**
        check DR_{C-1}, DR_{C-2}, etc to find a dose regimen which is not an **overdose** (F).
        if found, set $DR_C = DR_F$ **and** go to 2.
        else if not found, end trial with no MTD.
    end if
end if

* If in any case a dose regimen does not exist (i.e. $C - 1 < 1$ or $C + 1 > 7$), end trial with no MTD.
** If in any case 2 would require the recruitment of > 10 subjects to a single dose regimen, end trial with no MTD.

For each dose regimen DR_j, for each scenario, RJAGS was used to draw 10000 samples of $P(\text{non-CRS}|DR_j)$ using Eq. (8) and Gaussian priors with $\mu_{\alpha 2}, \mu_{\delta 2}, \sigma_{\alpha 2}, \sigma_{\delta 2}$ generated assuming $P(\text{non-CRS}|DR_1) = 0.05$ and $P(\text{non-CRS}|DR_7) = 0.2$. For each stage i, for each dose regimen DR_j, for each scenario, 10000 samples were drawn from $P(\text{CRS}_i|DR_j)$ using Eq. (7) and Gaussian priors on $[\alpha_{1i}], [\beta_{1i}], [\gamma_{1i}], i \in \{1, ..., S\}$ generated assuming $P(\text{CRS}|DR_4) = 0.1$ and $P(\text{CRS}|DR_7) = 0.3$ with ϵ set to 0.01.

To simulate whether, for a single patient, a CRS DLT occurred at a specific stage or a non-CRS DLT occurred at any point during a regimen's administration, a single sample of $P(\text{non-CRS}|DR_j)$ or $P(\text{CRS}_i|DR_j)$ was drawn uniformly at random. If the sample was $>= 0.5$, DLT occurred, if < 0.5, no DLT occurred. If a CRS DLT occurs at an early stage (i.e. $i < 5$), the subject is assumed to have withdrawn from the study, and is not counted for later stages when updating the posterior.

The standard BLRM was simulated using both the steady state dose from each regimen j as D_j, and the cumulative dose across each regimen j as D_j. These gave very similar results: for clarity only the results from the cumulative dose approach are given. For the priors to match between the standard BLRM and the extended BLRM, the extended BLRM prior defined above was used to estimate the overall probability of DLT at DR1 and DR7 (using Eq. 9), and then these estimates were used to generate point estimates of the means of the priors on α and β for the standard BLRM, as described in Sect. 2.2.1.

5 Results

Table 2 shows the operating characteristics for each of the three scenarios described, for both the standard BLRM and the modified BLRM. These are the percentage of trials which either stop with no MTD declared, or which declare an MTD with

Table 2 Operating Characteristics of standard BLRM versus CRS BLRM in three scenarios: where the prior beliefs about toxicity are correct, where the toxicity is higher than expected, and where the toxicity is as expected for the first few doses, but then increases much more rapidly than predicted. **Bold** indicates preferable behaviour, when the two models are compared

	% Trials declaring an MTD with true DLT rate in						% Trials stopped	
	Underdose		Target dose		Overdose			
Scenario	BLRM	CRS BLRM	BLRM	CRS BLRM	BLRM	CRS BLRM	BLRM	CRS BLRM
Priors correct	36.8	**15.5**	55.5	**73.5**	**2.2**	9.9	2.6	**1.1**
High toxicity	49.1	**8.0**	39.4	**86.1**	**5.4**	5.7	5.1	**0.1**
Low-high toxicity	61.0	**17.0**	21.2	**72.9**	15.2	**10.1**	1.6	**0.0**

the true DLT rate for that dose regimen in either the overdose (0.33–1], target dose (0.16–0.33], or underdose [0–0.16] ranges.

In all three scenarios, the modified BLRM correctly finds the target dose more often than the standard BLRM—a predicted success rate of over 70% in each case. The modified BLRM also declares an underdose as MTD less often than the standard BLRM in each scenario.

However the modified BLRM declares an overly toxic dose MTD more often than the standard BLRM both when the priors are correct and in the high toxicity scenario. It is less easily tricked than the standard BLRM by the low-high scenario however: the standard BLRM struggles in this setting as it tends to either declare MTD in a toxic setting or de-escalate aggressively and declare MTD with true DLT rate in the underdose range.

Overall, these results show that the modified BLRM is better suited than the standard BLRM to modelling the nuance of a setting where we have to escalate between dose regimens.

5.1 Real World Use

In a real phase I trial, the BLRM will be used to inform the decisions of a dose-escalation committee (DEC). This is a decision-making body which meets after each dose regimen has been tested in a small cohort of subjects to determine whether the dose regimen should be escalated, expanded, or de-escalated, or whether the trial should discontinued due to safety concerns.

Both the classic BLRM and our modified variant can be updated at each escalation step of the trial, given the data observed, to provide a full posterior over the dose-response relationship. This means that they can provide estimates of both the probability of overdose at the next potential dose level, and any intermediate dose level, allowing the DEC flexibility in its decisions.

One additional advantage of the modified BLRM is that it allows us to dig more deeply into this posterior: we can split by CRS DLTs versus non-CRS DLTs, and examine the effect of changing specific parts of the regimen on the total posterior. Figure 2 shows the contribution to the total posterior for the third regimen from each pre-steady state regimen stage, for CRS DLTs specifically. This level of granularity provides the DEC with the capacity to interrogate alternative regimens with relative ease.

As an example of why this capacity may be useful, consider DR5 in Table 1. This regimen includes a relative increase of 700 μg at the start of stage 4 (Day 14)—a much larger relative increase than any examined previously. If the first patient cohort to receive DR5 experience 2 CRS DLTs at stage 4, and no other DLTs have been observed in the study, a reasonable question might be: can we reduce the increase to 500 μg? Or, what is the maximum increase at stage 4 which would not lead to overdose? The modified BLRM has unmatched potential to support a DEC faced with this sort of scenario. Given phase I trials rely on well-informed decision making in

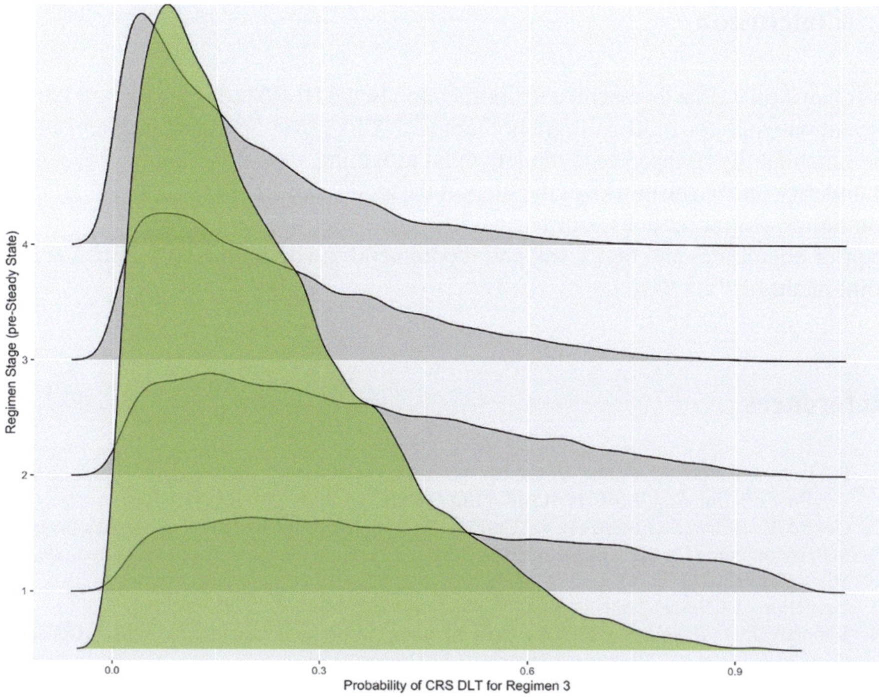

Fig. 2 Posterior densities (grey) contributing to overall posterior probability (green) of CRS for DR3

the face of minimal data, providing the kind of clarity demonstrated here can only be a good thing.

A main weakness of the modified BLRM is the substantial increase in parameters—particularly given we are in a setting where we have very minimal data. This makes prior elicitation harder, and sparse data even sparser (by modelling the probability of CRS occurrence at the level of stages, we are arguably diluting the effect of a CRS DLT on the whole model, compared to the simple BLRM).

However, we think that current approaches are simply not well suited to the problem of modelling a dose regimen in the context of dose-finding. In addition, because the modified BLRM is used flexibly to support decision making, rather than simply codifying a set of rules to follow (such as the 3+3 design, and similar), we think that the advantage of capturing the relationship between dose, prior exposure, and probability of CRS DLT outweighs the problems introduced by a more complex model.

6 Conclusion

In conclusion we have presented a modification to the BLRM for use in phase I dose-escalation trials for modern immunotherapies. This novel approach provides all of the functionality of the classic model whilst providing a substantial improvement in granularity, and hence improves the support the model can provide to a dose escalation committee or similar body. Further work is underway to explore better approaches to prior elicitation, and better ways to recommend the optimal next regimen at any point in a trial.

References

1. Babb, J., Rogatko, A., Zacks, S: Cancer phase I clinical trials: efficient dose escalation with overdose control. Stat. Med. **17**, 1103–1120 (1998)
2. Gerard, E., Zohar, S., Lorenzato, C., Ursino, M., Riviere, M.: Bayesian modeling of a bivariate toxicity outcome for early phase oncology trials evaluating dose regimens. Stat. Med. **40**(23), 5096–5114 (2021)
3. Guideline on Clinical Trials in Small Populations. European Medicines Agency (2006)
4. Kennedy, L.B., Salama, A.K.S.: A review of cancer immunotherapy toxicity. CA A Cancer J. Clin. **70**, 86–104 (2020). https://doi.org/10.3322/caac.21596
5. Le Tourneau, C., Lee, J.J., Siu, L.L.: Dose escalation methods in phase I cancer clinical trials. J. Natl. Cancer Inst. **101**(10), 708–720 (2009). https://doi.org/10.1093/jnci/djp079
6. Neuenschwander, B., Branson, M., Gsponer, T.: Critical aspects of the Bayesian approach to phase I cancer trials. Stat. Med. **27**(13), 2420–2439 (2008). https://doi.org/10.1002/sim.3230
7. Rogatko, A., Schoeneck, D., Jonas, W., Tighiouart, M., Khuri, F.R., Porter, A: Translation of innovative designs into phase I trials. J. Clin. Oncol. **25**(31), 4982–4986 (2007)
8. Yan, F., Mandrekar, S.J., Yuan, Y.: Keyboard: a novel Bayesian toxicity probability interval design for phase I clinical trials. Clin. Cancer Res. **23**(15), 3994–4003 (2017)
9. Araujo, D.V., Oliva, M., Li, K., Fazelzad, R., Liu, Z.A., Siu, L.L.: Contemporary dose-escalation methods for early phase studies in the immunotherapeutics era. Eur. J. Cancer **158**, 85–98 (2021)
10. Yuan, Y., Hess, K.R., Hilsenbeck, S.G., Gilbert, M.R.: Bayesian optimal interval design: a simple and well-performing design for phase I oncology trials. Clin. Cancer Res. **22**(17), 4291–4301 (2016)
11. Ananthakrishnan, R., Lin, R., He, C., Chen, Y., Li, D., LaValley, M.: An overview of the BOIN design and its current extensions for novel early-phase oncology trials. Contemp. Clin. Trials Commun. **28**, 100943 (2022)
12. Ji, Y., Liu, P., Li, Y., Bekele, B.N.: A modified toxicity probability interval method for dose-finding trials. Clin. Trials **7**(6), 653–63 (2010)
13. Guo, W., Wang, S.J., Yang, S., Lynn, H., Ji, Y.: A Bayesian interval dose-finding design addressing Ockham's razor: mTPI-2. Contemp. Clin. Trials **58**, 23–33 (2017)
14. Guo, X., Kent, S., Maity, A., Zhong, W.: Optimization of EWOC principle in BLRM design for phase 1 oncology trials. J. Biopharm. Stat. 1–17 (2024)

A Bayesian Multiple Ising Model

Alejandra Avalos-Pacheco, Andrea Lazzerini, Monia Lupparelli, and Francesco Claudio Stingo

Abstract Graphical models are a powerful tool for visually representing conditional independence structures of a set of variables. Recently, multiple graphical models for Gaussian variables have been extensively studied to analyse data coming from subgroups or subpopulations. However, proposals for binary sampling models remain limited. Here we present a methodological framework for Bayesian inference and model selection in multiple Ising models. We aim to model the variability introduced into a collection of binary variables due to external factors. The proposed Bayesian approach leverages conjugate priors and Laplace approximations, facilitating efficient model selection through a Metropolis-Hastings algorithm. Our methodological contributions are learning subgroup network structures for both model selection and parameter inference. We compare the performance of our proposed Bayesian method and other competing approaches, and show that our proposed method has a good performance in identifying related groups while offering balanced network sparsity and edge selection.

Keywords Binary data · Diaconis-Ylvisaker prior · Undirected graphs

A. Avalos-Pacheco (✉)
Johannes Kepler University Linz, Linz, Austria
e-mail: alejandra.avalos_pacheco@jku.at

Harvard-MIT Center for Regulatory Science, Boston, MA, USA

A. Lazzerini · M. Lupparelli · F. C. Stingo
University of Florence, Florence, Italy
e-mail: monia.lupparelli@unifi.it

F. C. Stingo
e-mail: francescoclaudio.stingo@unifi.it

A. Avalos-Pacheco et al. (eds.), *New Trends in Bayesian Statistics*,
Springer Proceedings in Mathematics & Statistics 511,
https://doi.org/10.1007/978-3-031-99009-0_4

1 Introduction

Graphical models have proven to be a highly popular tool for visualizing and understanding the conditional independence structures inherent in a collection of variables. Methods for estimating multiple graphical models, which aim to analyse data originating from distinct subgroups or subpopulations, have recently garnered substantial attention.

Among these, models tailored for Gaussian variables have been widely studied, in both the frequentist [1] and the Bayesian [2, 3] paradigms. However, the development of analogous models for binary random variables has been comparatively limited. In this paper, we present a methodological framework designed for Bayesian inference and model selection, a Bayesian multiple Ising model.

The Ising model [4] has been widely used in several research areas ranging from physics, for instance in the lattice gas [5], magnetism [4] and spin glass [6] models, to statistics in Potts models [7], graphical models [8], Markov Random Fields [9], among others.

Our focus lies in developing a Bayesian multiple Ising model (BMIM) able to capture the intricate variability introduced into a set of binary variables by external influences. A primary aim of our methodology is to uncover the underlying network structures characterizing various subgroups within the data. Additionally, we seek to provide both model selection and parameter inference for our proposed BMIM. Leveraging Bayesian principles, our proposed approach integrates Diaconis and Ylvisaker prior distribution [10], due to its conjugacy, and a Bernoulli prior. Inference is done via Laplace approximations, thereby facilitating efficient model selection through the Metropolis-Hastings algorithm. We remark that our current focus is on establishing a method that uses the full likelihood for posterior inference without encouraging a common structure across different groups. This serves as a foundational approach that can be extended in future work to incorporate information sharing. For multiple-Ising models that explicitly encourage a common structure across groups, allowing for information sharing to improve inference for heterogeneous networks, please refer to [15].

We provide a comprehensive evaluation of our proposed Bayesian methodology against two state-of-the-art alternative approaches: the Indep-Seplogit (SL) [11] and DataShared-SepLogit (DSSL) [12]. Through comparative analyses, we demonstrate the efficacy of our method in accurately identifying the graph structures for relevant subgroups while simultaneously achieving desirable properties such as network sparsity and precise edge selection.

By presenting a rigorous methodology for Bayesian inference and model selection in the context of multiple Ising models, we contribute to the broader landscape of graphical modeling techniques. Our findings underscore the importance of principled Bayesian approaches in addressing complex data analysis tasks, particularly in scenarios characterized by binary variables and subgroup heterogeneity.

The outline of this paper is as follows. Section 2 introduces our proposed Bayesian multiple Ising model and proposes prior formulations. Section 4 evaluates the performance of our model through simulations and comparisons with existing models. Section 5 concludes and provides venues for future work.

2 A Bayesian Multiple Ising Model (BMIM)

Let X be a categorical factor, taking the value $x \in \mathcal{X}$, $|\mathcal{X}| = q$. Let $G_{V|\mathcal{X}} = [G(x)]_{x\in\mathcal{X}}$ be a collection of multiple undirected graphs. Each graph $G(x) = (V, E(x))$ is defined by a set V of p nodes and a set $E(x)$ of edges which links pairs of nodes and depends on x, $x \in \mathcal{X}$. For any couple $r, j \in V$ and $x \in \mathcal{X}$, if $(r, j) \in E(x)$ there is an edge between r and j in the corresponding graph $G(x)$ while if $(r, j) \notin E(x)$ then the two nodes are disjointed in $G(x)$.

Let $Z_V|\{X = x\}$, be a p-dimensional binary random vector, indexed by the vertex set V, such that if $(r, j) \notin E(x)$ then $Z_r \perp\!\!\!\perp Z_j|\{Z_{V\setminus\{r,j\}}, X = x\}$. We assume that

$$Z_V|\{X = x\} \sim \text{Ising}(\lambda_x) \tag{1}$$

where $\lambda_x = [\lambda_{rj,x}]_{r,j\in V} \in \mathbb{R}^{p+(p\times(p-1))/2}$ is the canonical parameter [13]. Missing edges in each undirected graph $G(x)$ correspond to zero pairwise log-linear interactions, since $Z_r \perp\!\!\!\perp Z_j|\{Z_{V\setminus r,j}, X = x\}$ if and only if $\lambda_{rj,x} = 0$ in the related Ising(λ_x) model, for any $x \in \mathcal{X}$ [14].

We set the Diaconis and Ylvisaker prior distribution [10], as this is a conjugate prior for $\lambda_x = [\lambda_{rj,x}]_{r,j\in V} \in \mathbb{R}^{p+(p\times(p-1))/2}$, thus facilitating inference.

$$\lambda_x \sim \text{Diaconis-Ylvisaker}(\delta_x). \tag{2}$$

We finally set an independent Bernoulli prior on each element $\delta_x = [\delta_{rj,x}]_{r,j\in V} \in \{0, 1\}$

$$\delta_{rj,x} \sim \text{Bernoulli}(\pi_{rj,x}). \tag{3}$$

These selection parameters serve as a proxy that indicate which $\lambda_{rj,x}$ are significantly non-zero ($\delta_{rj,x} = 1$) and which are zero ($\delta_{rj,x} = 0$).

3 MCMC Algorithm

Posterior inference is conducted following the strategy in [15, 16]. Algorithm 1 summarizes the proposed MCMC algorithm.

Their proposed algorithm first update the graphs $G(x)$ for each x. It is important to notice that to compute the marginal likelihood $m(Z|G(x))$ and the posterior of λ_x,

Algorithm 1 MCMC Algorithm

Input: Binary data Z_V and categorical factors x
Output: Updated graphs $G(x)$ and samples of λ_x and δ_x
while $t < T$ **do**
  for each x **do**
    Sample λ_x and δ_x from their conditional posterior distribution
    Update the graph $G(x)$
    Perform a between-model and a within-model move
  end for
end while

it is required to compute the normalizing constants. As these normalizing constants are analytically intractable, they are calculated via a Laplace approximation [17].

Next, a Metropolis-Hastings algorithm is used to sample the λ_x and δ_x from their conditional posterior distributions. Given that the number of possible graphs grows exponentially with p^2, a stochastic model search is employed. For $x \in \mathcal{X}$, we start from $G(x)$ the graph accepted at the previous iteration and propose a new graph $\widetilde{G}(x)$ by randomly sampling one element of δ_x and switching its value. We then accept the new model $\widetilde{G}(x)$ with probability

$$r = \min\left(1, \frac{p(\widetilde{G}(x) \mid Z)}{p(G(x) \mid Z)}\right). \tag{4}$$

Finally, a between-model and a within-model move is done to improve mixing, as in [15, 16]. The stopping criteria is reaching a maximum number of iterations T. By default we set $T = 50,000$ with half of the chains as warm up.

We acknowledge that our implementation, while useful, is computationally tractable only for moderate values of p. The simulation studies presented in the following section, which considered $q = 4$ levels and $p = 10$ variables each, required an average of four hours to complete. Extensions to more efficient algorithms remain for future research.

4 Simulation Studies

We conduct a simulation study to assess the performance of our approach. We simulated data considering $q = 4$ levels from $Z_V|\{X = x\} \sim \text{Ising}(\lambda_x)$. Each level $x \in \mathcal{X} = \{0, 1, 2, 3\}$ consisted of 100 observations with associated undirected graph $\mathcal{G}$ with $p = 10$ nodes. Note that each graph includes at most $p(p-1)/2 = 45$ edges that are identified by the selection parameter δ_x, $x \in \mathcal{X}$. Here the graphs $G(0)$, $G(1)$ are identical to each other, but this pair is completely different from the graphs $G(2)$, $G(3)$, which are also identical to each other, as shown in Fig. 1. The graph structures $G(x)$ used here, are scale-free graphs using a preferential attachment

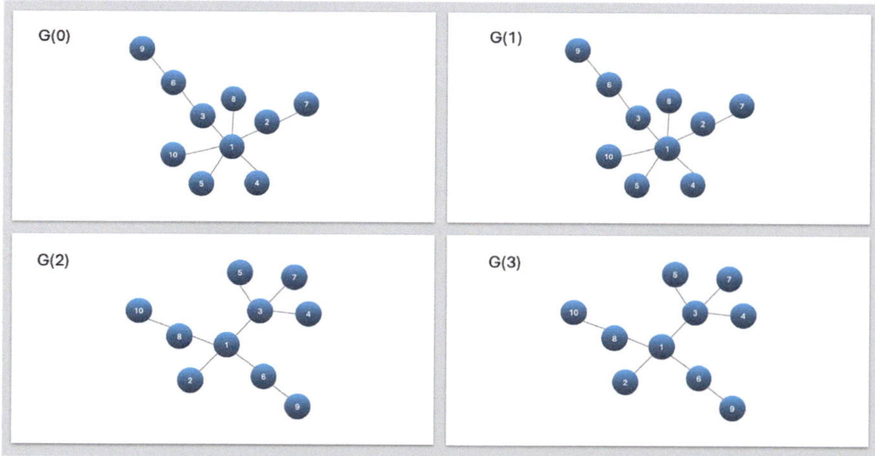

Fig. 1 Multiple undirected graphs in the simulated scenario

mechanism, leveraging the Barabási-Albert model to grow the network [18]. Graphs structures were generated using the `igraph` **R**-package [19].

In order to evaluate the performance of our proposed approach, we compare it with the Indep-Seplogit (SL) [11] and DataShared-SepLogit (DSSL) [12]. Both methods were implemented leveraging the `glmnet` **R**-package [20].

Estimation accuracy is measured using the area under the curve (AUC), accuracy, sensitivity and specificity between the true non-zero elements of λ_x, for all $x \in \mathcal{X}$ and the reconstruction obtained with all the methods.

Figure 2 provides the AUC and accuracy across ten different simulations. The results showed that, overall, the BMIM approach had a better performance than the SL and the DSSL, as they are based on approximate likelihood inference, showing the advantages of the exact likelihood method when there is strong homogeneity among the graphs. Furthermore, the DSSL displayed poor performance as it considered the graph homogeneity with very little flexibility, imposing similarities between groups a priori.

5 Conclusions and Future Directions

In this work, we showed the potential of a simple Bayesian multiple Ising model (BMIM) to learn subgroup network structures for binary data. Our BMIM makes use of the Diaconis-Ylvisaker conjugate prior for the canonical parameter and the Bernoulli prior for selection parameters. Inference is done via Laplace approximations, facilitating efficient model selection through a Metropolis-Hastings algorithm. Our simulation results show that our BMIM provides overall a higher performance

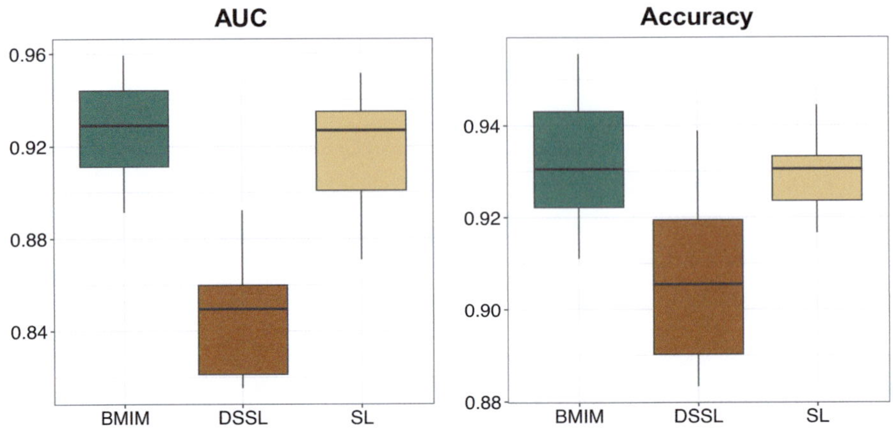

Fig. 2 AUC and accuracy for BMIM, DSSL and SL

that state of the art competitors. However, the simple model we have provided only hints at the true potential of BMIMs to efficiently model the conditional structures of a set of binary data.

Potential applications of our method include analysing binary survey data from different sub-populations, modelling binary economic decisions of various agents, and tracking the spread of diseases in diverse populations, where nodes represent individuals who are either infected or not. These examples highlight the versatility and practical relevance of our proposed method.

The likelihood based inference on the canonical parameter λ_x used here, is computationally tractable only for moderate values of p, because calculation of the normalizing constant requires computing a sum that grows exponentially in p. Those we aim to extend the work presented here to methods that approximate the likelihood, such as quasi-likelihood methods as in [15], or generalized Bayes methods leveraging a discrete Fisher divergence as in [21], or piecewise linear approximations for Approximate Bayesian Computation as in [22, 23].

Furthermore, the BMIM presented here do not allow linking the estimation of each group-specific graph to encourage common structures, which may lead to increased accuracy; Thus we aim to develop models that learn the common sparsity structures while allowing for group heterogeneities, leveraging hierarchical models or Markov Random Field priors as in [15, 24].

Finally the current BMIM does not employ any sparsity-inducing priors that may provide more interpretable networks. To this end we will build from efficient spike-and-slab priors [25] to learn the shared sparsity structures across different levels of the risk factors. We will use non-local spike-and-slab priors [26, 27], non-local based priors [28, 29], which yield model selection consistency properties.

Funding All the authors were partially supported by the Italian "Ministero dell'Universitá e della Ricerca", grant no. PRIN 2022SMNNKY. ML and FCS were partially supported by the "Dipartimenti Eccellenti 2023–2027" ministerial funds (Italy).

Competing Interests The authors have no conflicts of interest to declare that are relevant to the content of this chapter.

References

1. Danaher, P., Wang, P., Witten, D.M.: The joint graphical lasso for inverse covariance estimation across multiple classes. J. R. Stat. Soc. Ser. B **76**, 373–397 (2014)
2. Ha, M.J., Stingo, F.C., Baladandayuthapani, V.: Bayesian structure learning in multi-layered genomic networks. J. Am. Stat. Assoc. **116**(534), 605–618 (2021)
3. Peterson, C.B., Stingo, F.C., Vannucci, M.: Bayesian inference of multiple Gaussian graphical models. J. Am. Stat. Assoc. **110**(509), 159–174 (2015)
4. Ising, E.: Beitrag zur theorie des ferromagnetismus. Zeitschrift Für Physik A Hadrons and Nuclei **31**, 253–258 (1925)
5. Shi, Y., Duke, T.: Cooperative model of bacterial sensing. Phys. Rev. E **58**(5), 6399–6406 (1998)
6. Wang, J.S., Selke, W., Andreichenko, V.B., Dotsenko, V.S.: The critical behaviour of the two-dimensional dilute model. Phys. A **164**(2), 221–239 (1990)
7. Wu, F.Y.: The Potts model. Rev. Mod. Phys. **54**(1), 235–268 (1982)
8. Deshpande, A., Garofalakis, M., Jordan, M.: Efficient stepwise selection in decomposable models. In: Conference on Uncertainty in Artificial Intelligence (UAI) (2001)
9. Sherrington, D., Kirkpatrick, S.: Solvable model of a spin-glass. Phys. Rev. Lett. **35**(35), 1792–1796 (1975)
10. Diaconis, P., Ylvisaker, D.: Conjugate priors for exponential families. Ann. Stat. **7**(2), 269–281 (1979)
11. Meinshausen, N., Bühlmann, P.: High-dimensional graphs and variable selection with the Lasso. Ann. Stat. **34**(3), 1436–1462 (2006)
12. Ollier, E., Viallon, V.: Regression modelling on stratified data with the lasso. Biometrika **104**(1), 83–96 (2017)
13. Barndorff-Nielsen, O.E.: Information and Exponential Families in Statistical Theory. Wiley, New York (1978)
14. Whittaker, J.: Graphical Models in Applied Multivariate Analysis. John Wiley & Sons, Chichester (1990)
15. Avalos-Pacheco, A., Lazzerini, A., Lupparelli, M., Stingo, F.C.: Bayesian inference of multiple Ising models for heterogeneous public opinion survey networks. J. R. Stat. Soc. Ser. C Appl. Stat. [Advanced Publication] (2025)
16. Lazzerini, A.: Profile graphical models. Ph.D. thesis, University of Bologna (2021)
17. Tierney, L., Kadane, J.P.: Accurate approximations for posterior moments and marginal densities. J. Am. Stat. Assoc. **81**(393), 82–86 (1986)
18. Albert, R., Barabási, A.-L.: Statistical mechanics of complex networks. Rev. Mod. Phys. **74**(1), 47–97 (2002)
19. Csardi, G., Nepusz, T.: The igraph software package for complex network research. Inter J. Complex Syst. 1695 (2006)
20. Jerome, F., Trevor, H., Robert, T.: Regularization paths for generalized linear models via coordinate descent. J. Stat. Softw. **33**(1), 1–22 (2010)
21. Matsubara, T., Knoblauch, J., Briol, F.-X., Oates, C.J.: Generalised Bayesian inference for discrete intractable likelihood. J. Am. Stat. Assoc. [advance publication] (2023)
22. Moores, M.T., Drovandi, C.C., Mengersen, K., Robert, C.P.: Pre-processing for approximate Bayesian computation in image analysis. Stat. Comput. **25**(1), 23–33 (2015)

23. Moores, M., Nicholls, G.K., Pettit, A.N., Mengersen, K.: Scalable Bayesian inference for the inverse temperature of a hidden Potts model. Bayesian Anal. **15**(1), 1–27 (2020)
24. Peterson, C.B., Stingo, F.C., Vannucci, M.: Bayesian inference of multiple Gaussian graphical models. J. Am. Stat. Assoc. **110**(509), 159–174 (2015)
25. George, E.I., McCulloch, R.E.: Approaches for Bayesian variable selection. Statistica Sinica 339—374 (1997)
26. Avalos Pacheco, A.: Factor regression for dimensionality reduction and data integration techniques with applications to cancer data. Ph.D. thesis, University of Warwick (2019)
27. Avalos-Pacheco, A., Rossell, D., Savage, R.E.: Heterogeneous large datasets integration using Bayesian factor regression. Bayesian Anal. **17**(1), 33–66 (2022)
28. Johnson, V.E., Rossell, D.: On the use of non-local prior densities in Bayesian hypothesis tests. J. R. Stat. Soc. Ser. B Stat. Methodol. **72**(2), 143–170 (2010)
29. Johnson, V.E., Rossell, D.: Bayesian model selection in high-dimensional settings. J. Am. Stat. Assoc. **107**(498), 649–660 (2012)

Bayesian Nonparametric Estimation of Time-Varying Macroeconomic Tail Risk

Ramsés H. Mena, Matteo Ruggiero, and Amandeep Singh

Abstract We propose a Bayesian nonparametric approach for assessing macroeconomic tail risk using a time-dependent Dirichlet process mixture model. The model is applied to a datasets that covers several decades across both OECD and non-OECD countries, and is shown to captures fluctuations in extremely negative outcomes in consumption level changes. The analysis reveals left-skewed distributions consistent with the association of downside risks, and allows to evaluate these aspects dynamically and in comparison to upside risks. Such insights are invaluable for policymakers and investors, offering nuanced perspectives to navigate shifting economic landscapes and implement tailored risk management strategies effectively.

Keywords Dependent Dirichlet process · Gibbs sampler · GDP

1 Introduction

Macroeconomic tail risk pertains to the probability of rare and severe economic events like financial crises or recessions, which impact consumption, investment, and overall economic well-being. Understanding macroeconomic tail risk is crucial for several reasons. Policymakers use tail risk metrics to assess the opportunity of policy interventions towards greater economic stability, and integrating these metrics into economic forecasting enhances forecast accuracy, aiding decision-making in both public and private sectors. In asset pricing, it helps explain anomalies such

R. H. Mena
IIMAS-UNAM, Mexico City, Mexico
e-mail: ramses@sigma.iimas.unam.mx

M. Ruggiero
University of Torino and Collegio Carlo Alberto, Turin, Italy
e-mail: matteo.ruggiero@unito.it

A. Singh (✉)
University of Torino, Turin, Italy
e-mail: amandeep.singh263@edu.unito.it

A. Avalos-Pacheco et al. (eds.), *New Trends in Bayesian Statistics*,
Springer Proceedings in Mathematics & Statistics 511,
https://doi.org/10.1007/978-3-031-99009-0_5

as the equity premium puzzle, guiding investment decisions by incorporating risk preferences. For investors and financial institutions, managing this risk is essential for portfolio allocation and risk mitigation. Overall, comprehending and quantifying macroeconomic tail risk is essential for guiding investment decisions, managing risks, and formulating effective economic policies.

We address the estimation of macroeconomic tail risk using a Bayesian nonparametric approach through a time dependent mixture model. More specifically, we use a semiparametric mixture of parametric kernels, with dynamic mixing measure given by a discrete-time dependent Dirichlet process.

2 Dataset

We consider a dataset initially constructed by [1], which consists of real per capita consumption expenditures and gross domestic product (GDP) levels. The sample covers 42 countries, for many of which the data are available since the early nineteenth century. Indeed, the data span from 1870 to 2020 and comprise 25 countries in the Organization for Economic Cooperation and Development (OECD), 14 countries from Latin America and Asia, as well as Egypt, Russia, and South Africa. Figure 1 shows the time series for OECD and non-OECD countries.

As macroeconomic pitfalls generally unfold slowly over time, we will primarily work with consumption levels by taking a time window of H years over which levels are measured. The variable of interest is the ratio of consumption levels measured H years apart, namely

$$c_{i,t}^{(H)} := \frac{C_{i,t}}{C_{i,t-H}},$$

where $C_{i,t}$ is the *per capita* real consumption recorded in year t for country i. We set $H = 3$ to have a three-year temporal window for all countries considered and drop the H superscript henceforth.

Crisis events are commonly characterized by a decline in consumption levels, often defined as falling below a specific threshold relative to previous levels, such as one standard deviation below the long-term mean. We will denote this occurrence as a Crisis Event of type A. More severe crises are identified when growth rates plummet below two standard deviations from the long-term mean. We will designate the latter event as a Crisis Event of type B. Although rare events in statistics are typically related to farther ends of the distribution tails, these parameters make sense from a macroeconomic perspective, see [4] for further details.

The crisis event is therefore defined as

$$\text{Crisis}_{i,t}(\kappa) := \begin{cases} 1, & \text{if } c_{i,t} < \text{mean}(c_{i,\cdot}) - \kappa \times \text{SD}(c_{i,\cdot}) \\ 0, & \text{otherwise} \end{cases} \tag{1}$$

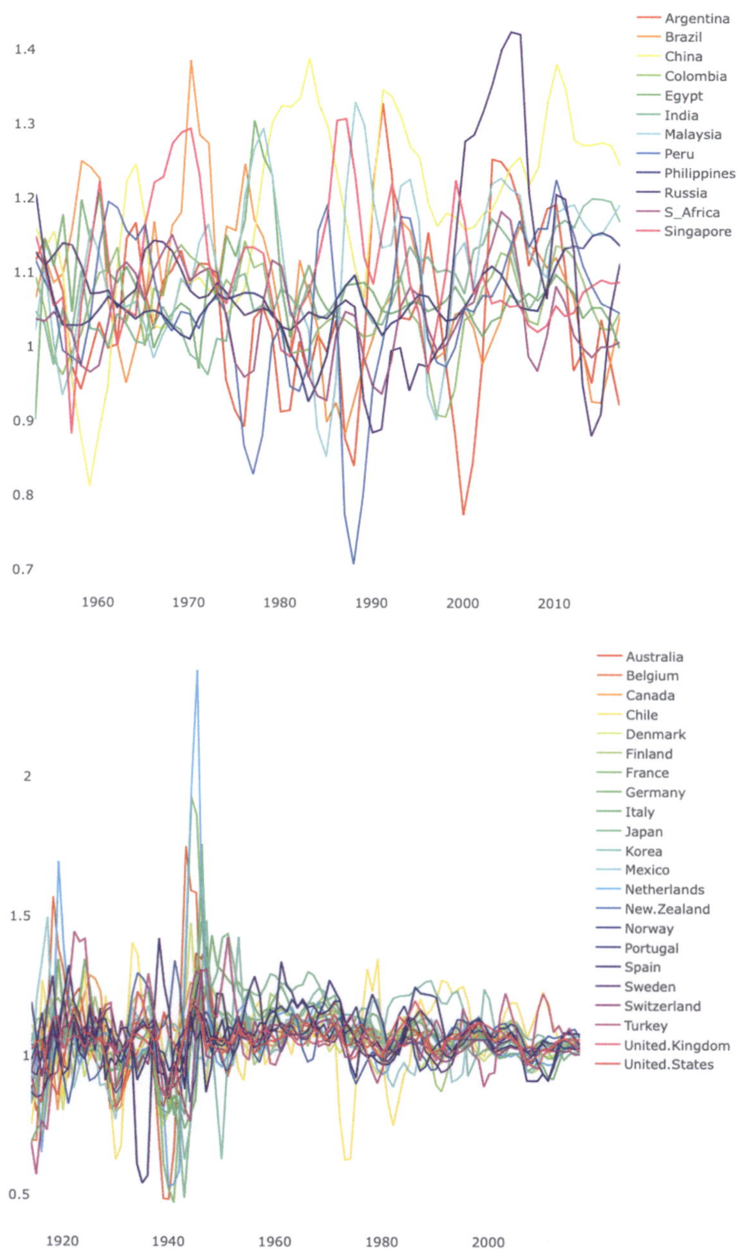

Fig. 1 Time series plot of consumption ratios for non-OECD (top) and OECD (bottom) countries

where mean$(c_{i,\cdot})$ is the mean of $\{c_{i,t}, t = 1, \ldots, T\}$, and similarly for the standard deviation SD$(c_{i,\cdot})$. Setting $\kappa = 1, 2$ defines a crisis event of type A or B respectively.

Note that according to the above definition, a type B event is also a type A event. However, to avoid unnecessary complications, we ignore this redundancy in terms of notation, and consider as type A events those that satisfy (1) for $k = 1$ but not for $k = 2$.

3 A Dependent Dirichlet Process Mixture

We consider a discrete time model, with temporal indices $\{t_m, m = 1, \ldots, M\}$ replacing the role of t in the previous section. We model the variable of interest c_{i,t_m} through a time-indexed random density function $f_{P_{t_m}} : \mathcal{S} \to \mathbb{R}^+$, defined by the mixture model

$$f_{P_{t_m}}(y) = \int_{\Theta} K(y|\theta) P_{t_m}(d\theta), \quad y \in \mathcal{S}, \tag{2}$$

where y is the state variable for c_{i,t_m}. Here $\mathcal{S}$ is the sample space, $K(\cdot|\theta)$ is a kernel density on $(\mathcal{S}, \mathcal{B}(\mathcal{S}))$, $\theta \in \Theta \subset \mathbb{R}^d$, and $\{P_{t_m}, m = 1, \ldots, M\}$ is a collection of dependent random probability measures on $(\Theta, \mathcal{B}(\Theta))$.

Functionals of interest are here the mean functional

$$M_{t_m} = \int_{\mathcal{S}} y f_{P_{t_m}}(dy),$$

and the survival function associated to $f_{P_{t_m}}$, which measures the probability of exceeding a threshold y_0, defined as

$$X_{t_m}(y_0) := \mathbb{P}_{t_m}(y > y_0) = \int_{y_0}^{\infty} f_{P_{t_m}}(y)dy.$$

We let the time-indexed random probability measure P_{t_m} in (2) be a dependent Dirichlet process (DDP) (see [5] for a recent review). Specifically, we adopt the choice with time-varying weights and fixed atoms proposed in [2] (cf. also [3]), whereby

$$P_{t_m} = \sum_{j=1}^{\infty} w_j(t_m)\delta_{\theta_j}, \quad \theta_j \overset{\text{i.i.d.}}{\sim} F_0, \tag{3}$$

with $\sum_{j=1}^{\infty} w_j(t_m) = 1$ almost surely for all $m = 1, \ldots, M$, and where F_0 is a non-atomic probability measure on Θ. The weights $w_j(t_m)$ are obtained through stick-breaking, where $\{w_j(t_m), m = 1, \ldots, M\}$ is a realization of the stochastic process $\{W_j(t_m), m = 1, \ldots, M\}$, where

$$W_j(t_m) = V_j(t_m) \prod_{i=1}^{j-1}(1 - V_i(t_m)), \qquad j \geq 1, \tag{4}$$

and each *stick-breaking component* $\{V_j(t_m), m = 1, \ldots, M\}$ is a Markov chain. The latter is in turn set to start from $V_j(t_1) \sim Q$, with $Q := \text{Beta}(1, \vartheta)$ for $\vartheta > 0$, with transition probabilities defined, for a given $\varepsilon \in (0, 1)$, as

$$\mathbb{P}(V_j(t_{m+1}) \in A | V_j(t_m) = v) = (1 - \varepsilon)Q(A) + \varepsilon\delta_v(A), \quad m = 2, \ldots, T. \tag{5}$$

See [2]. It is easy to see that the single stick-breaking components $V_j(t)$ are updated after geometric holding times, independently across $j \geq 1$, with a value drawn from their stationary distribution Beta$(1, \vartheta)$. It is also easy to see that P_{t_m} is marginally a Dirichlet process, and ε controls the autocorrelation of each stick-breaking component, thus (indirectly) also the autocorrelation of the time-varying process $P = \{P_{t_m}, m = 1, \ldots, M\}$. Extremes cases are given by the asymptotic regime $\varepsilon \to 0$, in which case all stick breaking components are resampled independently from Q with probability one at every time step, resulting in P_{t+1} being independent from P_t for every t, and $\varepsilon \to 1$, in which case each stick-breaking component V_j is fixed through time, resulting in no temporal evolution for P_t, which equals its initial state P_0 at every t.

To complete the specification of the DDP model for the application we have in mind, we select a kernel $K(\mathbf{y}|\theta) := \log N(\mathbf{y}|\mu, \sigma^2)$, where $\log N$ is a univariate log-Normal distribution. The centering measure is set to be

$$F_0(\mu, \sigma^2) := N(\mu|\mu_0, \sigma_0^2)\text{IG}(\sigma^2|\alpha, \beta) \tag{6}$$

where IG$(\cdot|\alpha, \beta)$ denotes the inverse gamma distribution with expected value given by $\beta/(\alpha - 1)$ for $\alpha > 1$.

As a result, the full Bayesian model is

$$\begin{aligned} y|\mu, \sigma^2 &\sim \log N(\mathbf{y}|\mu, \sigma^2) \\ (\mu, \sigma^2)|\mathbf{w}(t), (\mu_j, \sigma_j^2) &\sim \sum_{j=1}^{\infty} w_j(t)\delta_{(\mu_j, \sigma_j^2)}, \\ (\mu_j, \sigma_j^2)|\mu_0, \sigma_0^2, \alpha, \beta &\overset{\text{i.i.d.}}{\sim} N(\mu_j|\mu_0, \sigma_0^2)\text{IG}(\sigma_j^2|\alpha, \beta) \end{aligned}$$

where $\{w_j(t_m), m = 1, \ldots, M\}$ evolve as in (4) and (5), and $\mu_0, \sigma_0^2, \alpha, \beta$ are hyperparameters. As a consequence, the observations are treated here as partially exchangeable, given at every t_m we have an exchangeable model, but the dynamics of the stick-breaking components induce correlation among the models at different time points.

4 Posterior Inference

Posterior inference is performed via Gibbs sampler with slice sampling steps to handle each random probability measure in the DDP. Following [2], the Gibbs sampler is based on the data augmentation obtained through the joint density

$$f_{P_{t_m}}(y, u, s) = I(u < w_s(t_m))K(y|\theta_s), \tag{7}$$

where s identifies the mixture component and u is a uniform random variable typical of slice samplers. It is easy to check that integrating out s and u yields P_{t_m} as expressed in (2).

Let us now assume we observe n trajectories at times $(t_1, t_2, \ldots, t_T)$, for some fixed T. For $i = 1, 2 \ldots n$ and $m = 1, 2 \ldots T$, the associated augmented likelihood is

$$L_{\mathbf{v},\boldsymbol{\theta}}(\mathbf{y}, \mathbf{u}, \mathbf{s}) = \prod_{i=1}^{n} \prod_{m=1}^{T} I(u_{i,m} < w_{s_{i,m}}(t_m))K(y_{i,m}|\theta_{s_{i,m}}), \tag{8}$$

where $\mathbf{y}, \mathbf{u}, \mathbf{s}$ are now vectors, $\mathbf{v} := \{(v_{1,m}, v_{2,m}, \ldots); m = 1, \ldots T\}$ denotes the infinite collection of state-time observations of the stick-breaking components in (4), i.e. $\{V_j(t_m) = v_{j,m}\}$, and $\boldsymbol{\theta} := (\theta_1, \theta_2, \ldots)$ is the infinite set of random locations. The Gibbs sampler algorithm alternates updating $v_{j,m}, s_{i,m}, u_{i,m}$ and θ_j, together with the stick-breaking components according to (5). We refer the reader to [2] for further details on the algorithm.

5 Application

We first apply the model to a set of non-OECD countries with data spanning from 1953 to 2019. We set hyperpriors $\vartheta \sim \text{Ga}(a, b)$ and $\varepsilon \sim \text{Beta}(\gamma, \nu)$, with $a = 1$, $b = 1$, $\gamma = 2$, and $\nu = 2$. We also set the remaining hyperparameters in (6) to be $\mu_0 = 0$, $\sigma_\mu^2 = 0.01$, $\alpha = 300$, $\beta = 1$. This prior specification provides essentially noninformative distributions, particularly for μ and ϑ. The centering measure is set to be $F_0(\mu, \sigma^2) := N(\mu|0, 0.01)\text{IG}(\sigma^2|300, 1)$.

We implement the Gibbs sampler with 50K iterations, of which the first 20K as burn-in, and thinning every 30 iterations for a total of 1000 samples.

Figure 2 shows the marginal density estimates (black solid) obtained through the dynamic model for two single years, namely 1987 (left) and 2006 (right), together with the pointwise 95% credible intervals obtained by taking, for each abscissa, the quantiles over the time series of density estimates (red and green dotted curves). The peculiar shape of the green curve is an effect of two periods of relative stability along the time series, each giving rise to a representative shape for that period. The distribution in 1987 is within the credible region for what concerns the right tail, while the left tail shows considerably more mass than most of the time series, being outside the credible region. The distribution in 2006 has considerable mass in the right tail, indicating an outlier probability of growth in the consumption levels. The credible regions per se show that modes to the right of 1, indicating economies that have a general tendency to grow.

Figure 3 shows the posterior probabilities of A and B Crisis Events respectively, where values exceeding 0.2 and 0.1 respectively are highlighted.

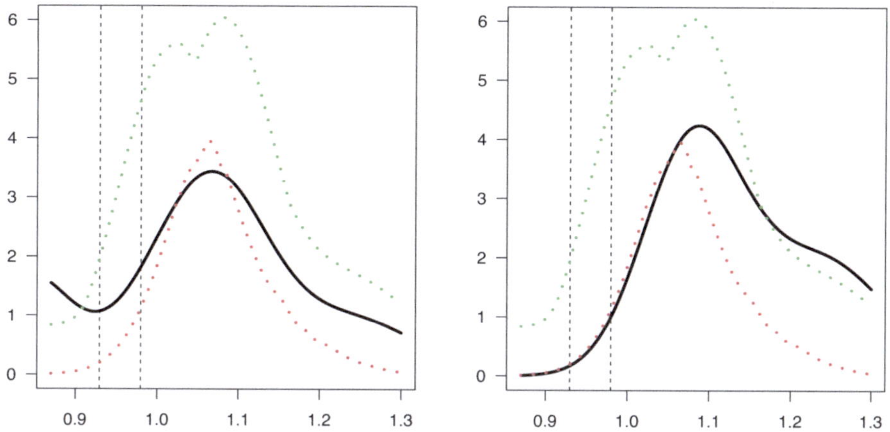

Fig. 2 Marginal density (solid black) for non-OECD countries in years 1987 (left) and 2006 (right). The red and green dotted curves show the pointwise 95% credible intervals computed for each abscissa over the entire time series of density estimates. The vertical lines are the long-term thresholds used to identify crises events

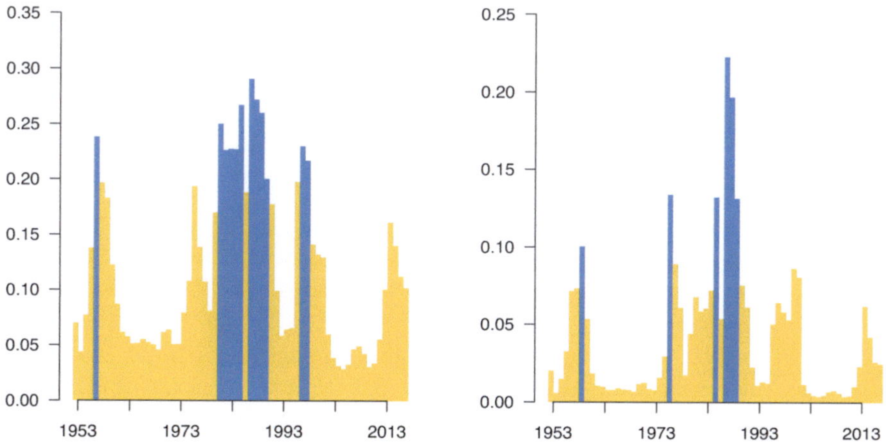

Fig. 3 Crisis probabilities for event A (left) and B (right) for different years, with highlighted values greater than 0.2 (left) and 0.1 (right)

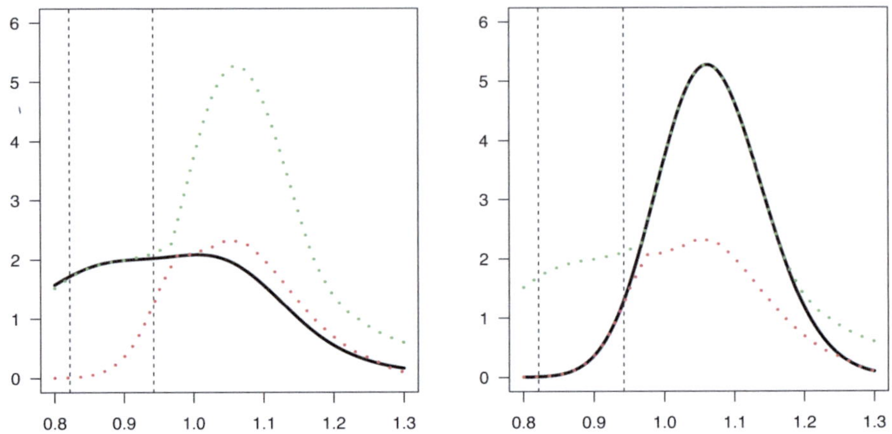

Fig. 4 Marginal density (solid black) for OECD countries in years 1940 (left) and 2016 (right). The red and green dotted curves show the pointwise 95% credible intervals computed for each abscissa over the entire time series of density estimates. The vertical lines are the long-term thresholds used to identify crises events

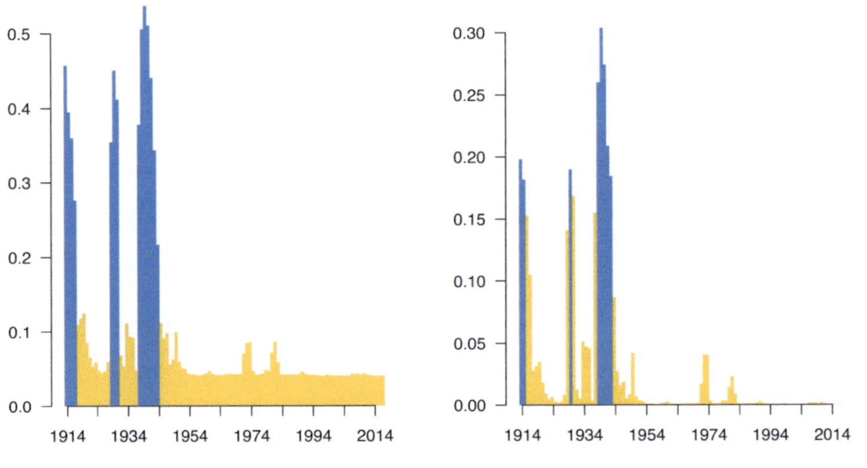

Fig. 5 Crisis probabilities for event A (left) and B (right) for different years, with highlighted values greater than 0.2 (left) and 0.17 (right)

Next we analyze OECD countries with data spanning from 1914 to 2019, with analogous prior specification. Figures 4 and 5 have analogous interpretation to Figs. 2 and 3. The generally slimmer credible regions, as compared to Fig. 2, are consistent with the lower variability shown by OECD countries in Fig. 1. In particular, the year 1940 (left panel of Fig. 4) exhibits very heavy left tail, which coincides with the upper 95% pointwise quantiles, indicating a high probability of crisis events. The tail asymmetry emphasizes the nature of downside risk compared to upside risk, as recessions are typically associated with left-skewed distributions, highlighting the need to monitor these quantities over time. The year 2016 (right panel of Fig. 4), on the contrary, exhibits a central region of the curve coinciding with the upper quantiles and tails coinciding the lower quantiles. This is due to the fact that 2016 belongs in a group of very similar distributions which end up determining the pointwise 95% quantiles for the credibility region. Such group of years is characterized by great economic stability with the lowest probability of sharp consumption changes (in both directions) and the highest probability of moderate economic growth (the mode being slightly to the right of 1).

6 Discussion

A Bayesian nonparametric approach based on dependent Dirichlet processes proves a very flexible approach for suitably capturing nonlinear features within the analyzed time-series, typically presenting greater adaptability than parametric alternatives. It also exhibits improved capability in accommodating correlation across time, avoids the need to impose homoscedasticity assumptions, and can account for heavy-tailedness. The application to the current dataset reveals a notable temporal fluctuations in the left tail of the consumption ratio distribution. This aspect can be observed in both Figs. 2 and 4, where the amplitude of the credible region concerning the left tails is a result of these fluctuations over time. These clearly indicate great variability in the probability of extreme negative outcomes, such as economic downturns or crises, across different time periods. Grasping the dynamics of these time-varying tail risks holds paramount significance for policymakers and investors alike, enabling them to tailor strategies and policies effectively to evolving economic landscapes.

Following a similar intuition, a comparison of Figs. 2 and 4 reveals a very different degree of temporal variability for the distributions of interest between OECD and non-OECD countries, somehow providing empirical evidence of a well known fact in macroeconomic dynamics.

Furthermore, the observation that the right tail of the distribution, representing positive economic dynamics, remains relatively steady compared to the left tail, at least for OECD countries, underscores an inherent imbalance in economic outcomes. This can be seen for example in Fig. 4, where the credible region in correspondence of the left tail is substantially fatter than the right counterpart. This asymmetry implies greater likelihood of extreme negative events occurring compared to extreme positive events, commonly termed as "risk asymmetry". Recognizing this imbalance is critical for ensuring that risk management practices adequately factor in the potential ramifications of adverse economic shocks.

References

1. Barro, R.J., Ursua, J.F.: Macroeconomic data. Dataset available at https://scholar.harvard.edu/barro/data_sets (2010)
2. Gutierrez, L., Mena, R.H., Ruggiero, M.: A time dependent Bayesian nonparametric model for air quality analysis. Comput. Stat. Data Anal. **95**, 161–175 (2016)
3. Gutierrez, L., Mena, R.H., Ruggiero, M.: On GEM diffusive mixtures. In: JSM Proceedings, Section on Nonparametric Statistics. American Statistical Association, Alexandria, VA (2016)
4. Marfè, R., Pénasse, J.: Measuring macroeconomic tail risk. J. Financ. Econ. **156**, 103838 (2024)
5. Quintana, F.A., Müller, P., Jara, A., MacEachern, S.N.: The dependent Dirichlet process and related models. Stat. Sci. **37**, 24–41 (2020)

A Metropolis–Hastings Algorithm for Sampling Coagulated Partitions

Marco Dalla Pria, Matteo Ruggiero, and Dario Spanò

Abstract Motivated by privacy constraints which force to unlabel multiple samples from a discrete distribution, we consider the problem of dealing with a Bayesian model when conditioning on multiple partitions induced by these samples. Aiming at evaluating the distribution of coagulations produced by matching the groups of the multiple partitions, whose original type is unknown, and motivated by the high computational cost of exact evaluations, we formulate a Metropolis–Hastings sampler that is shown to yield good approximations in reasonable computing time despite the great sparsity displayed by the target distribution.

Keywords Ewens sampling formula · Unlabeled data · Coagulation of partitions

1 Introduction

Priors which almost surely select discrete distributions are widely used in Bayesian nonparametric modelling and inference; see, e.g., [3, 4]. Observations $Y_1, \ldots, Y_n$ drawn from a random discrete distribution will feature ties with positive probability. Dropping the observation labels induces a random partition of n

$$\gamma = (\gamma_1, \ldots, \gamma_J) : \ \gamma_j \in \mathbb{N}, \ \gamma_1 \geq \gamma_2 \geq \cdots \geq \gamma_J > 0, \ |\gamma| := \sum_{j=1}^{J} \gamma_j = n,$$

M. Dalla Pria (✉)
University of Torino, Turin, Italy
e-mail: marco.dallapria@unito.it

M. Ruggiero
New York University Abu Dhabi, Abu Dhabi, UAE
e-mail: mr7690@nyu.edu

D. Spanò
University of Warwick, Coventry, England
e-mail: d.spano@warwick.ac.uk

A. Avalos-Pacheco et al. (eds.), *New Trends in Bayesian Statistics*,
Springer Proceedings in Mathematics & Statistics 511,
https://doi.org/10.1007/978-3-031-99009-0_6

where γ_j is the j-th *part* after ranking decreasingly by size, and J the number of parts. The interpretation is that γ_j is the absolute frequency of the j-th most popular distinct value (type) appearing in the sample $Y_1, \ldots, Y_n$.

Here we are interested in the case when M such n-sized samples $Y^{(m)} = (Y_1^{(m)}, \ldots, Y_n^{(m)})$, $m = 1, \ldots, M$, are collected separately from the same population, but only their respective induced partitions $\gamma^{(m)} = (\gamma_1^{(m)}, \ldots, \gamma_{J(m)}^{(m)})$, $m = 1, \ldots, M$, are available. Namely, the data have been unlabeled so the *type* associated to each individual or group of ties is unknown, and the only information available is the vector of group sizes for each of the M datasets. In particular, this determines that the correspondence between pair of groups $\gamma_j^{(m)}$ and $\gamma_h^{(m')}$ is not available either, i.e., we do not know whether these groups correspond to the same or to different types. This setting can arise when multiple samples are collected by different entities and the type labels have been removed for privacy issues or other technical reasons, and is further motivated by a filtering problem for infinite-dimensional unlabeled diffusion models that the authors are developing elsewhere.

Given such a framework, any statistical problem of interest goes through the evaluation of the probability of a specific *coagulation*, which is the partition induced by a specific *matching* of the M original partitions. These notions are defined in the next section, and can be informally thought of as first assigning (latent) labels to the partition parts, in such a way that within each original partition no two groups share the same label, then summing the group sizes associated to the same label and finally ranking decreasingly. The coagulation is then the partition of nM items whose part sizes are sums of parts of the original partitions.

2 Partition Matching and Coagulation

Denote by $\gamma^{(1:M)} = (\gamma^{(1)}, \ldots, \gamma^{(M)})$ the collection of M partitions defined as above. We refer to the component $\gamma_j^{(m)}$ as *part* j of partition m. A *matching* of $\gamma^{(1:M)}$ is an arrangement in nonempty *groups* of the $\sum_{m=1}^{M} J(m)$ parts of $\gamma^{(1:M)}$, treated as distinguishable objects, such that each group contains at most one element from each original partition. See e.g. [5, 6] and references therein for an overview on matching problems and the MCMC approach to tackle them. For example, if $\gamma^{(1)} = (\gamma_1^{(1)})$, $\gamma^{(2)} = (\gamma_1^{(2)}, \gamma_2^{(2)})$, namely we have observed one type $\gamma_1^{(1)}$ times in the first dataset, and two distinct types, respectively $\gamma_1^{(2)}$ and $\gamma_2^{(2)}$ times, in the second dataset, then the possible matchings are

$$\{\{\gamma_1^{(1)}\}, \{\gamma_1^{(2)}\}, \{\gamma_2^{(2)}\}\}, \quad \{\{\gamma_1^{(1)}, \gamma_1^{(2)}\}, \{\gamma_2^{(2)}\}\}, \quad \{\{\gamma_1^{(1)}, \gamma_2^{(2)}\}, \{\gamma_1^{(2)}\}\}. \tag{1}$$

In the leftmost matching in the above example, the three parts are considered all different, whereas in the second and third $\gamma_1^{(1)}$ is considered as of the same type as $\gamma_1^{(2)}$ or $\gamma_2^{(2)}$. Each matching induces a *coagulation* of $\gamma^{(1:M)}$, that is a partition of the nM items obtained by considering the group sizes resulting from the matching, ranked in decreasing order. If in (1) we have $n = 3$ and $\gamma^{(1)} = (3),\ \gamma^{(2)} = (2, 1)$, the coagulations we obtain are respectively

$$(3, 2, 1), \quad (5, 1), \quad (4, 2).$$

Note that that different matchings can induce the same coagulation. For example, if $n = 2$ and $\gamma^{(1)} = (2),\ \gamma^{(2)} = (1, 1)$, we obtain

$$(2, 1, 1), \quad (3, 1),$$

where $(3, 1)$ is obtained in two ways, from $\{\{\gamma_1^{(1)}, \gamma_1^{(2)}\}, \{\gamma_2^{(2)}\}\}$ and $\{\{\gamma_1^{(1)}, \gamma_2^{(2)}\}, \{\gamma_1^{(2)}\}\}$.

Denote by $\mathrm{Coag}(\gamma^{(1:M)})$ the set of all the coagulations that can be obtained from all the matchings of $\gamma^{(1:M)}$, and by $\kappa(\mu|\gamma^{(1:M)})$ the number of matchings of $\gamma^{(1:M)}$ that yield the specific coagulation $\mu \in \mathrm{Coag}(\gamma^{(1:M)})$. Denote also a matching

$$\mathcal{M} = \{\mathcal{M}_1, \ldots, \mathcal{M}_{K(\mathcal{M})}\}$$

where $\mathcal{M}_k$ is a subset of parts $\{\gamma_{j_1}^{(m_1)}, \ldots, \gamma_{j_{d_k}}^{(m_{d_k})}\}$ with $1 \leq m_1 \neq \cdots \neq m_{d_k} \leq M$, so that $\mathcal{M}_k$ contains at most one part from each $\gamma^{(m)}$.

The matching group sizes are

$$\mu_k = \sum_{j,m:\ \gamma_j^{(m)} \in \mathcal{M}_k} \gamma_j^{(m)}, \quad k = 1, \ldots, K(\mathcal{M})$$

and we let $\mu(\mathcal{M})$ be the descending arrangement of $(\mu_k)_{k=1,\ldots,K(\mathcal{M})}$, defined as the coagulation induced by $\mathcal{M}$.

3 The Probability of a Coagulation

We assume from now on that partitions are induced by Dirichlet populations, i.e., the law $\pi(\gamma)$ of a partition $\gamma = (\gamma_1, \ldots, \gamma_J)$ is described by the celebrated *Ewens sampling formula*. Cf. [3], Sect. 14.1.1. This can in turn be written as

$$\mathbb{P}_\theta(\gamma) := C(\gamma)\ \mathrm{EPPF}_\theta\,(\gamma)$$

where, for $|\gamma| = \sum_{j=1}^J \gamma_j$ and $a_i(\gamma) = \sum_{j=1}^J \mathbb{1}(\gamma_j = i)$,

$$C(\gamma) = \binom{|\gamma|}{\gamma_1, \ldots, \gamma_J} \frac{1}{a_1(\gamma)! \ldots a_{|\gamma|}(\gamma)!},$$

and, for $\theta > 0$, EPPF_θ denotes the *exchangeable partition probability function*

$$\text{EPPF}_\theta\,(\gamma) = \frac{\theta^J}{\theta_{(|\gamma|)}} \prod_{j=1}^{J} (\gamma_j - 1)!$$

where we have denoted by $a_{(b)} = a(a+1)\dots(a+b-1)$ the Pochhammer symbol or *rising factorial.*

The following result identifies the probability of a coagulation $\mu \in \text{Coag}(\gamma^{(1:M)})$.

Theorem 1 *For $\mu \in \text{Coag}(\gamma^{(1:M)})$ we have*

$$\mathbb{P}_\theta(\mu|\gamma^{(1:M)}) \propto \kappa(\mu|\gamma^{(1:M)})\ \text{EPPF}_\theta\,(\mu). \tag{2}$$

Proof The proof proceeds by induction on M. Let first $M = 2$. In light of the the well-known construction of EPPF_θ through a *Chinese restaurant process* (cf, e.g., [3], Sect. 14.1.1), consider a restaurant having n customers and table j having $\gamma_j^{(1)}$ customers. We first need the joint probability $\mathbb{P}_\theta(\mu, \gamma^{(2)}|\gamma^{(1)})$ that the next n customers sit at the restaurant as described by $\gamma^{(2)}$ and the final configuration of the restaurant is μ. For each matching $\mathcal{M}$ yielding μ, there are $C(\gamma^{(2)})$ ways to sit the n new customers. A matching $\mathcal{M}$ represents a configuration of the restaurant, whose conditional probability given $\gamma^{(1)}$ is

$$\frac{C(\gamma^{(2)})}{(\theta+n)_{(n)}} \prod_{\{\gamma_i^{(1)},\gamma_j^{(2)}\}\in\mathcal{M}} (\gamma_i^{(1)})_{(\gamma_j^{(2)})} \prod_{\{\gamma_j^{(2)}\}\in\mathcal{M}} \theta(\gamma_j^{(2)} - 1)!.$$

This expression is constant on matchings that yield the same coagulation μ, hence it equals

$$C(\gamma^{(2)})\frac{\text{EPPF}_\theta\,(\mu)}{\text{EPPF}_\theta\,(\gamma^{(1)})}.$$

By definition there are $\kappa(\mu|\gamma^{(1)}, \gamma^{(2)})$ of such matchings, and we have just proved that if $\mu \in \text{Coag}(\gamma^{(1:2)})$,

$$\mathbb{P}_\theta(\mu, \gamma^{(2)}|\gamma^{(1)}) = C(\gamma^{(2)})\kappa(\mu|\gamma^{(1)}, \gamma^{(2)})\frac{\text{EPPF}_\theta\,(\mu)}{\text{EPPF}_\theta\,(\gamma^{(1)})}. \tag{3}$$

Therefore, through (3), we have

$$\mathbb{P}_\theta(\mu|\gamma^{(1)}, \gamma^{(2)}) \propto \kappa(\mu|\gamma^{(1)}, \gamma^{(2)})\ \text{EPPF}_\theta\,(\mu).$$

Assume the result holds for $M-1$ partitions, and $\gamma^{(M)}$ is given. Then

$$\begin{aligned}\mathbb{P}_\theta(\mu|\gamma^{(1:M)}) &= \frac{\mathbb{P}_\theta(\mu,\gamma^{(M)}|\gamma^{(1:M-1)})}{\mathbb{P}_\theta(\gamma^{(M)}|\gamma^{(1:M-1)})}\\ &\propto \sum_{\lambda\in\mathrm{Coag}(\gamma^{(1:M-1)})} \mathbb{P}_\theta(\mu,\gamma^{(M)},\lambda|\gamma^{(1:M-1)})\\ &\propto \sum_{\lambda\in\mathrm{Coag}(\gamma^{(1:M-1)})} \mathbb{P}_\theta(\lambda|\gamma^{(1:M-1)})\mathbb{P}_\theta(\mu,\gamma^{(M)}|\lambda)\end{aligned}$$

where the last line follows from the conditional independence of $(\mu,\gamma^{(M)})$ and $\gamma^{(1:M-1)}$ given the configuration of the restaurant λ after the first $M-1$ partitions of customers have sit.

The induction hypothesis and the result for $M=2$ now yield

$$\begin{aligned}&\mathbb{P}_\theta\ (\mu|\gamma^{(1:M)})\\ &\propto C(\gamma^{(M)}) \sum_{\lambda\in\mathrm{Coag}(\gamma^{(1:M-1)})} \kappa(\lambda|\gamma^{(1:M-1)})\ \mathrm{EPPF}_\theta\ (\lambda)\kappa(\mu|\lambda,\gamma^{(M)})\frac{\mathrm{EPPF}_\theta\ (\mu)}{\mathrm{EPPF}_\theta\ (\lambda)}\\ &\propto \mathrm{EPPF}_\theta\ (\mu) \sum_{\lambda\in\mathrm{Coag}(\gamma^{(1:M-1)})} \kappa(\lambda|\gamma^{(1:M-1)})\kappa(\mu|\lambda,\gamma^{(M)})\end{aligned}$$

and noting that

$$\sum_{\lambda\in\mathrm{Coag}(\gamma^{(1:M-1)})} \kappa(\lambda|\gamma^{(1:M-1)})\kappa(\mu|\lambda,\gamma^{(M)}) = \kappa(\mu|\gamma^{(1:M)}),$$

leads to the result. □

4 A Metropolis–Hastings for Coagulated Partitions

While $\mathrm{EPPF}_\theta\ (\mu)$ is the model for the population, $\kappa(\mu|\gamma^{(1:M)})$ in (2) is a purely combinatorial object which does not have an explicit expression, which makes the distribution (2) unamenable for pointwise evaluation. In principle κ should be counted by explicitly listing the matchings and counting how many of these produce a given coagulation. However, the support of $\mathrm{Coag}(\gamma^{(1:M)})$ grows extremely fast, as shown for a very simple case in Table 1. In addition, Fig. 1 illustrates the execution times for the target distribution via a brute force approach. The increase of computation time is dramatic, as for example, for $\gamma^{(1)}=\cdots=\gamma^{(M)}=(4,3,2,1)$, the target was computed in less than a second for $M=2$ and in more than 3 h for $M=6$.

Table 1 Number of different coagulations derived from equal partitions $\gamma^{(m)} = (3, 2, 1)$, for $m = 1, \ldots, M$

	$M = 2$	$M = 3$	$M = 4$	$M = 5$	$M = 6$
$\lvert \mathrm{Coag}(\gamma^{(1:M)}) \rvert$	23	154	787	3189	11295

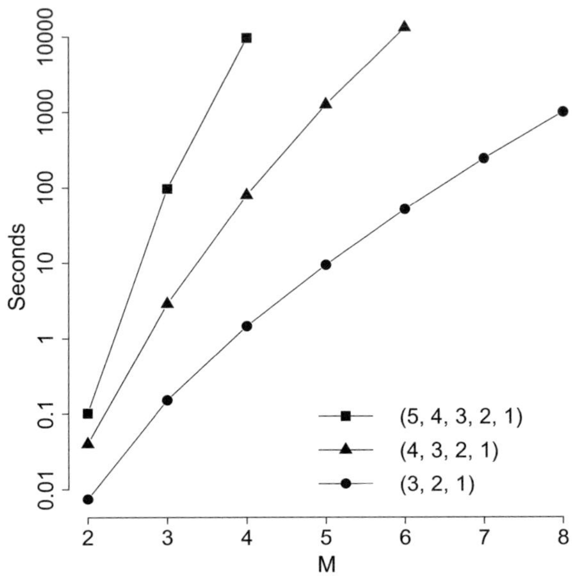

Fig. 1 Execution times (seconds in log scale) for the exact computation of the target distribution $\mathbb{P}_\theta(\mu|\gamma^{(1:M)})$ with a brute force approach as function of M, where $\theta = 1$, $\gamma^{(1)} = \cdots = \gamma^{(M)} = \gamma$, where γ alternatively equals $(3, 2, 1)$, $(4, 3, 2, 1)$ or $(5, 4, 3, 2, 1)$

Given the above example is produced with a very small sample size n for each $\gamma^{(m)}$, it is immediate to realise that this direct approach is unfeasible even for moderate sample sizes. We thus need to resort to simulation.

In the case $M = 2$, [2] proposed a way to simulate matchings uniformly, so that after collecting matchings that yield the same coagulation one obtains sample from a distribution proportional to $\kappa(\mu|\gamma^{(1:2)})$. This exact sampler could be exploited to obtain an estimate of (2), when used as a proposal distribution for Monte Carlo methods. It can be shown that using the uniform distribution over matchings as proposal for a self-normalized importance sampler, is equivalent to an EPPF_θ-tilting of the resulting empirical distribution over $\mathrm{Coag}(\gamma^{(1:2)})$. Despite this strategy is theoretically sound and the derived estimator of (2) is consistent (see e.g. [1]), due to the sparsity of (2) it may be computationally inefficient. This phenomenon is shown in Fig. 2 for a relatively simple case given partitions $\gamma^{(1)} = \gamma^{(2)} = \gamma^{(3)} = (5, 4, 3, 2, 1)$ and $\theta = 1$. The figure suggests that a Monte Carlo strategy based on a uniform proposal of matchings is likely to waste computational resources by proposing samples from areas of the support having negligible target probability.

Instead of aiming for an extension of [2], namely an exact sampler of matchings for the case $M \geq 2$ to be used as proposal in a self-normalised importance sampler,

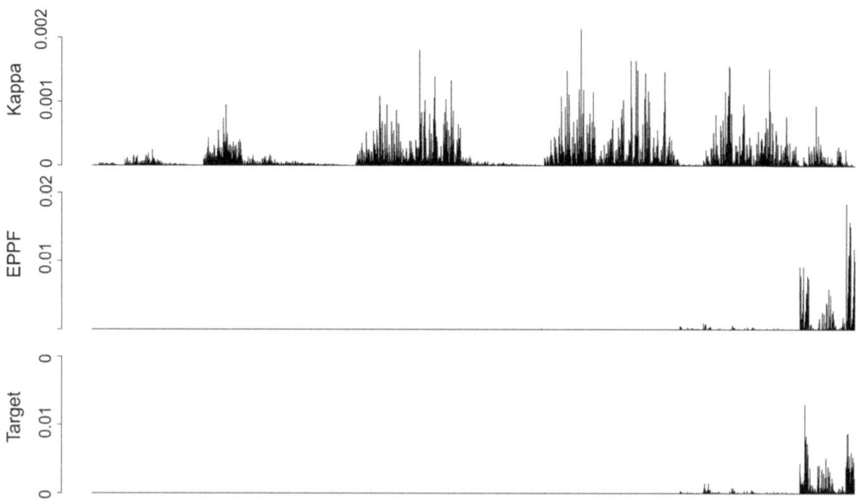

Fig. 2 Images based on partitions $\gamma^{(1)} = \gamma^{(2)} = \gamma^{(3)} = (5, 4, 3, 2, 1)$ and $\theta = 1$, which yield 17364 possible coagulations, sorted from the longest (leftmost point of the support) to the shortest (rightmost point of the support). From the top: normalised $\kappa(\mu|\gamma^{(1:3)})$; normalised EPPF_θ; target distribution, obtained by multiplying and normalising the top two panels. The figure highlights the high multimodality and sparsity of the considered distributions

we instead aim here at designing an ergodic Markov chain on the set of matchings that explores the most relevant regions with respect to the target distribution.

Through the induced coagulation, we can evaluate the EPPF_θ on matchings by $\text{EPPF}_\theta(\mathcal{M}) := \text{EPPF}_\theta(\mu(\mathcal{M}))$, so that to each matching we assign mass proportional to $\text{EPPF}_\theta(\mathcal{M})$. Now partition the set of all matchings via the equivalence relation whereby matchings in the same class induce the same coagulation, i.e., $\mathcal{M} \sim \mathcal{M}' \Leftrightarrow \mu(\mathcal{M}) = \mu(\mathcal{M}')$ and note that the equivalence classes are in one-to-one correspondence with $\text{Coag}(\gamma^{(1:M)})$. Hence this bijection induces a probability on $\text{Coag}(\gamma^{(1:M)})$ obtained by summing the probabilities of matchings in each equivalence class. Since in each equivalence class the probability is constant, in light of the definition of κ which counts the number of matchings that yield μ, the resulting distribution on $\text{Coag}(\gamma^{(1:M)})$ corresponds to (2). It is therefore enough to find a Markov chain on the space of matchings with stationary distribution proportional to $\text{EPPF}_\theta(\mathcal{M})$. Then at stationarity, once lumped onto coagulations, such Markov chain is in state $\mu \in \text{Coag}(\gamma^{(1:M)})$ with probability (2).

We restrict ourselves to transition kernels that move one partition part at a time, i.e. such that the transition probability $p(\mathcal{M}'|\mathcal{M})$ is positive if and only if either $\mathcal{M} = \mathcal{M}'$ or if there exists a unique $\gamma_j^{(m)}$ such that, neglecting a possible empty group,

$$\bigcup_{k=1}^{K(\mathcal{M})} \{\mathcal{M}_k \setminus \{\gamma_j^{(m)}\}\} = \bigcup_{k=1}^{K(\mathcal{M}')} \{\mathcal{M}'_k \setminus \{\gamma_j^{(m)}\}\} \tag{4}$$

i.e., removal of $\gamma_j^{(m)}$ gives identical matchings. We can provide such Markov chain in the form of a *Metropolis–Hastings*, where uniform proposals are followed by an acceptance step. The part $\gamma_j^{(m)}$ to be moved is chosen uniformly at random from $\{\gamma_j^{(m)} : 1 \leq m \leq M, 1 \leq j \leq J(m)\}$. The proposed matching is such that (4) holds, i.e., we propose a group assignment $\tilde{k}$ for $\gamma_j^{(m)}$, and the resulting matching $\tilde{\mathcal{M}}$ is accepted with probability $\min\{1, \text{EPPF}_\theta(\tilde{\mathcal{M}})/\text{EPPF}_\theta(\mathcal{M})\}$. The output $\mathcal{M}'$ of the above one-step transition, either equal to $\tilde{\mathcal{M}}$ or to $\mathcal{M}$, identifies the new matching, and its corresponding coagulation is $\mu(\mathcal{M}')$. The pseudocode for the above Metropolis–Hastings is outlined in Algorithm 1.

Algorithm 1: `MH transition`

Input : current state of the Markov chain $\mathcal{M} = \{\mathcal{M}_1, \ldots, \mathcal{M}_{K(\mathcal{M})}\}$
Output : next state $\mathcal{M}'$

choose a part $\gamma_j^{(m)}$ uniformly at random
$\tilde{\mathcal{M}} \leftarrow \bigcup_{k=1}^{K(\mathcal{M})}\{\mathcal{M}_k \setminus \{\gamma_j^{(m)}\}\}$
$\tilde{\mathcal{M}} \leftarrow \tilde{\mathcal{M}} \cup \{\varnothing\}$
choose $\tilde{\mathcal{M}}_{\tilde{k}} \in \tilde{\mathcal{M}}$ with uniform probability if $\gamma_{j'}^{(m)} \notin \tilde{\mathcal{M}}_{\tilde{k}}\ \forall j' \neq j$ and zero otherwise
$\tilde{\mathcal{M}}_{\tilde{k}} \leftarrow \tilde{\mathcal{M}}_{\tilde{k}} \cup \{\gamma_j^{(m)}\}$
$\tilde{\mathcal{M}} \leftarrow \tilde{\mathcal{M}} \setminus \{\varnothing\}$
$\mathcal{M}' \leftarrow \tilde{\mathcal{M}}$ with probability $\min\left\{1, \frac{\text{EPPF}_\theta(\tilde{\mathcal{M}})}{\text{EPPF}_\theta(\mathcal{M})}\right\}$, else $\mathcal{M}$
return $\mathcal{M}'$

5 Illustration

We illustrate the behaviour of the MCMC introduced above for small sample sizes, which allows to make a comparison with the exact distribution, as the latter becomes quickly too expensive to evaluate even for moderate sample sizes.

Figure 3 shows the exact distributions on the set of coagulated partitions that can be obtained from $\gamma^{(1)} = \gamma^{(2)} = \gamma^{(3)} = (3, 2, 1)$, when $\theta = 1$, proportional to $\kappa(\mu|\gamma^{(1:3)})$ (top panel) and corresponding to $\text{EPPF}_\theta(\mu)$ (center panel), and compares (bottom panel) the target (2) (solid) with the MCMC-based estimate (dashed). The simulation-based estimate of the target shows a good fit, despite the great sparsity of the target which puts almost zero mass in most of the support. Figure 4 provides a magnified version of the bottom panel of Fig. 3, limited to the part of the support where the target puts most of its mass. Similarly, the motivating example in Fig. 2 is tackled and illustrated in Fig. 5, which directly provides a zoomed-in view of the interesting part of the support.

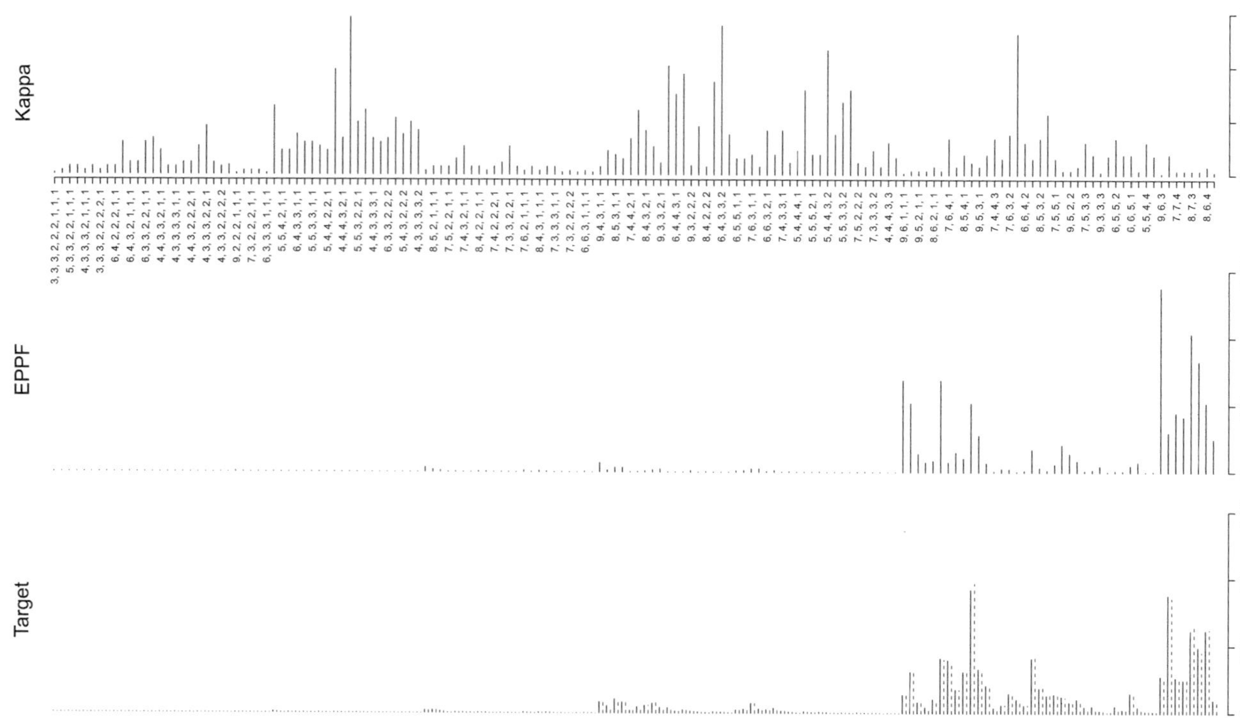

Fig. 3 Distributions on coagulations that can be obtained from $\gamma^{(1)} = \gamma^{(2)} = \gamma^{(3)} = (3, 2, 1), \theta = 1$. Coagulations $\mu \in \text{Coag}(\gamma^{(1:3)})$ are shown on the horizontal axis (only for the top panel). Top: counts $\kappa(\mu|\gamma^{(1:3)})$ of matchings that produce μ. Middle: $\text{EPPF}_\theta(\mu)$. Bottom: target (2) (solid black) vs. MCMC-based empirical estimate (dashed blue). The MCMC was run for 10^6 steps and started from the longest coagulation (3, 3, 3, 2, 2, 2, 1, 1, 1). The total variation distance between target and estimate resulted in 0.011

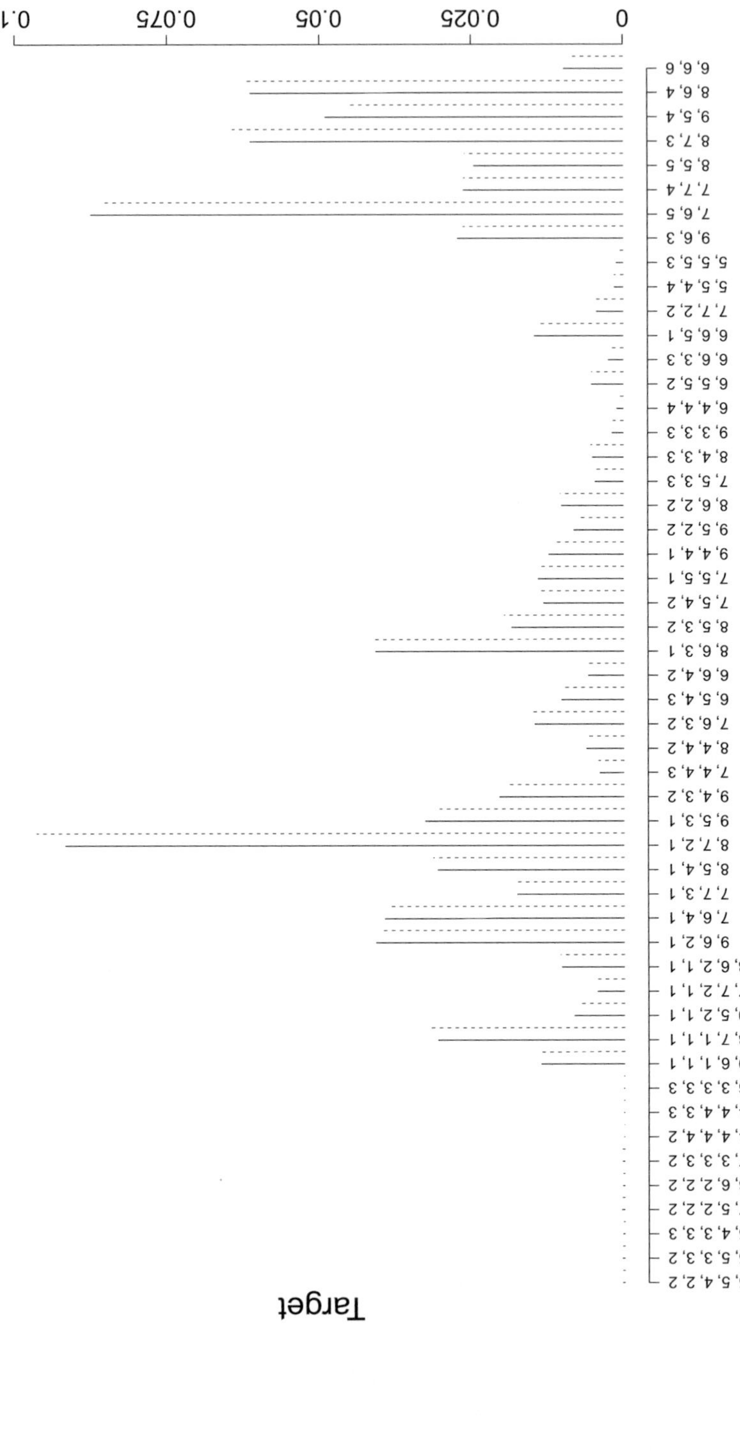

Fig. 4 Zoom on the bottom panel of Fig. 3, relative to the subset of the support where the target puts most of its mass. Target (solid black) vs. MCMC estimate (dashed blue)

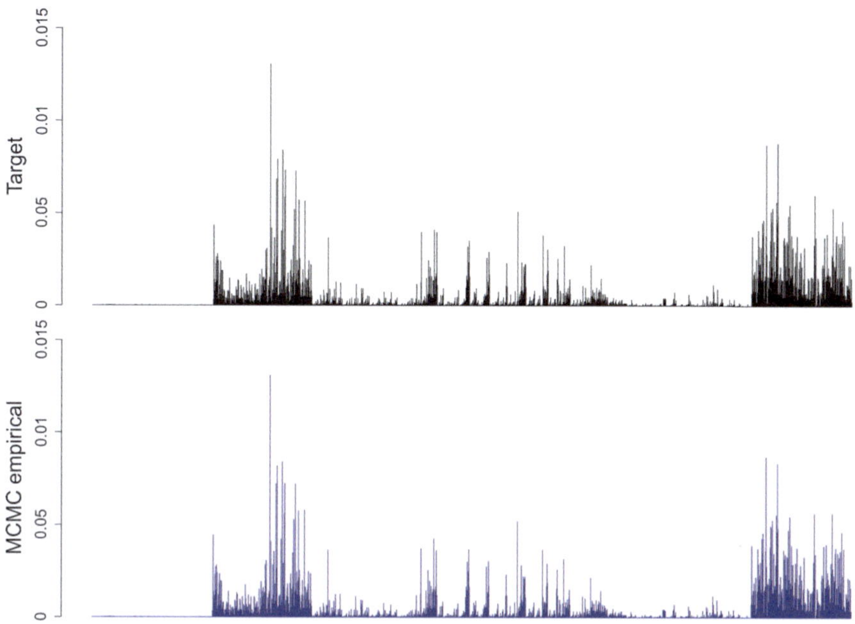

Fig. 5 Top: target zoomed on the 1500 rightmost coagulations of the support shown in Fig. 2. Bottom: MCMC-based empirical estimate. The MCMC was run for 10^7 steps and started from the coagulation (15, 12, 9, 6, 3). The total variation distance between target and estimate resulted in 0.033

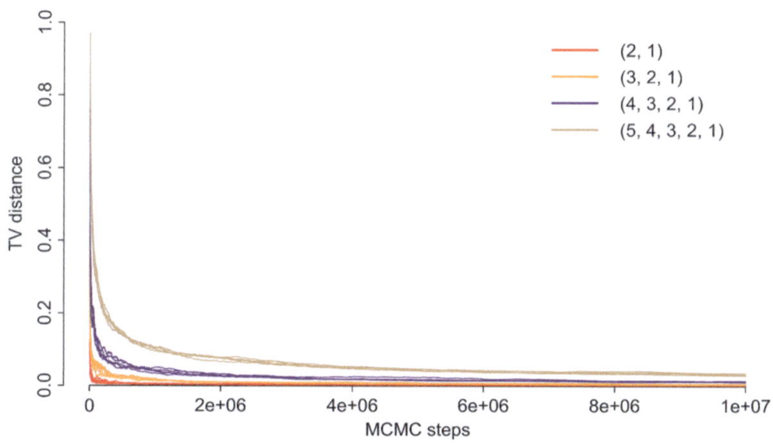

Fig. 6 Total variation distance every 10^3 steps between $\mathbb{P}_\theta(\mu|\gamma^{(1:M)})$ and the MCMC empirical for $\theta = 1$ and $\gamma^{(1)} = \gamma^{(2)} = \gamma^{(3)} = \gamma$, for $\gamma = (2, 1), (3, 2, 1), (4, 3, 2, 1)$ and $(5, 4, 3, 2, 1)$. Each chain was started from the longest coagulation, e.g. (2, 2, 2, 1, 1, 1), and 5 chains were run for each γ. Average runtime (minutes): 10.48 for (2, 1), 11.23 for (3, 2, 1), 12.11 for (4, 3, 2, 1) and 13.04 for (5, 4, 3, 2, 1)

Given most common diagnostics of convergence designed for real-valued chains cannot be used for chains of integer partitions, in Fig. 6 we provide empirical evidence of the convergence (with average runtimes) under some small-dimensional scenarios by looking at the total variation distance between the target, when this can be computed via brute force in an acceptable amount of time, and the MCMC empirical, as the number of steps increases.

References

1. Chopin, N., Papaspiliopoulos, O.: An Introduction to Sequential Monte Carlo, vol. 4. Springer (2020)
2. Dalla Pria, M., Ruggiero, M., Spanò, D.: Exact sampling of two coagulated partitions. In: Methodological and Applied Statistics and Demography. SIS 2024, Short Papers, Contributed Sessions. Springer, Cham (2024)
3. Ghosal, S., van der Vaart, A.: Fundamentals of Nonparametric Bayesian Inference. Cambridge Books (2017)
4. Hjort, N.L., Holmes, C., Müller, P., Walker, S.G.: Bayesian Nonparametrics, vol. 28. Cambridge University Press (2010)
5. Jerrum, M.: Counting, Sampling and Integrating: Algorithms and Complexity. Springer Science & Business Media (2003)
6. Sinclair, A.: Algorithms for Random Generation and Counting: a Markov Chain Approach. Springer Science & Business Media (2012)

Conditionally Partially Exchangeable Partitions for Dynamic Networks

Francesco Gaffi

Abstract In network analysis, understanding the dynamics of evolving networks is often of paramount importance. We introduce and study a novel class of models to detect evolving communities underlying dynamic network data. The methods build upon the established literature on stochastic block models and extend it to accommodate temporal evolution. The cornerstone of our approach is the interplay of random partitions induced by hierarchical normalized completely random measures and the assumption of conditional partial exchangeability, a recently introduced modeling principle for capturing the dynamic of evolving partitions within a Bayesian framework. Our methodology effectively addresses the limitations inherent in traditional static community detection methods, and in contrast with other dynamic extensions of the classical stochastic block models, provides flexibility and built-in uncertainty quantification, while inducing a form of distributional invariance coherent with a time-evolving clustering scheme.

Keywords Bayesian nonparametrics · Conditional partial exchangeability · Dynamic network · Random partition · Stochastic block model

1 Introduction

Stochastic block models (SMBs) are a foundational method within network analysis, entailing a probabilistic approach to capturing the underlying structure of complex networks. See [12] for an early account. SBMs assume that nodes in a network can be partitioned into distinct groups or "blocks", with nodes within the same block exhibiting similar connectivity behaviour, i.e. same connection probabilities with nodes from the other blocks. By estimating these connection probabilities, SBMs can efficiently represent various network topologies, stepping beyond the strong assumption that communities of nodes must present dense connections within

F. Gaffi (✉)
Department of Mathematics, University of Maryland, College Park, USA
e-mail: fgaffi@umd.edu

A. Avalos-Pacheco et al. (eds.), *New Trends in Bayesian Statistics*,
Springer Proceedings in Mathematics & Statistics 511,
https://doi.org/10.1007/978-3-031-99009-0_7

and sparse connections outside, entailed by other community detection methods. In fact, this simple but flexible modeling paradigm allows researchers to uncover complex latent community structures that often characterize real-world networks. See e.g. [1, 7, 15, 19, 22].

Traditional SBMs assume a static network setting, overlooking the temporal dynamics prevalent in many real-world scenarios. Hence, in the last decade, advancements in dynamic network analysis have led to the development of temporal extensions of SBMs, aiming to capture evolving community structures over time. These dynamic SBMs provide a powerful framework for understanding how networks change over different time intervals, thus offering a more comprehensive perspective on the intricate dynamics of interconnected systems. See, for instance, [3, 5, 10, 17, 18, 20, 23, 25–28]. However, except for a few contributions, such like [5, 23], where the underlying community structure is modeled via a Dirichlet process, these methods require fixing the number of communities underlying the network, thus, often lacking in flexibility and robustness. Moreover, the existing Bayesian proposals do not focus the attention on crafting prior distributions inducing a suitable form of probabilistic symmetry among the nodes across time. Often, this results in models that ignore part of the information conveyed by the network structure. See [4, 8] on this matter. In dynamic networks, where one observes the evolution of connections among the same set of nodes, it is crucial to not lose the *node identity* information, i.e. the fact that nodes across time represent the same statistical unit. In this article, we propose a novel class of Bayesian models for the detection of evolving communities within the framework of SBMs. The communities at each time point are modeled via the random partition induced by hierarchical normalized completely random measures, defined in [2], where they are called hierarchical normalized random measures with independent increments (H-NRMIs). Such comprehensive class of nonparametric priors generalizes the Dirichlet process and other Gibbs type priors to the partially exchangeable framework, as intended by de Finetti in [6]. More importantly, in our proposal temporal dependence is flexibly captured by means of the general assumption of conditional partial exchangeability, recently introduced in [9]. As a result, the temporal information flow is guided by an appropriate predictive distributional invariance, that, while granting flexibility, takes also into account that the network observed at different time points is indeed composed by the same set of nodes, preserving node identity. The present work constitutes part of a more extensive study of methodologies for clustering and prediction in dynamic networks, driven by appropriate distributional invariance assumptions, carried out in [8].

The rest of this paper is structured as follows. In Sect. 2.1, we review the classical SBMs and focus on partially exchangeable stochastic block models (pEx-SBMs), a recent innovative class of models introduced in [4], that constitute the static conditional structure of our novel dynamic model. Section 2.2 is dedicated to outlining conditional partial exchangeability, the distributional invariance that we employ to model the clustering information flow through time. We describe the model structure and summarize its features in Sect. 3. Finally, Sect. 4 is dedicated to the formulation of a marginal algorithm for posterior inference.

2 Background

2.1 Stochastic Block Models and Recent Extensions

A static undirected network with no self-loops, composed by V nodes indexed in $[V] = \{1, \ldots, V\}$, can be represented by a symmetric adjacency matrix $\mathbf{Y}$, which is a $V \times V$ matrix whose generic entry y_{uv} equals 1 when there exists an edge connecting nodes u and v and equals 0 otherwise. To represent the absence of self-loops, the diagonal of $\mathbf{Y}$ can be put to 0.

The classical SBM assumes the existence of a number H of unobservable exhaustive and mutually exclusive communities of nodes. So that, given the community allocations, the connections among nodes are modeled as independent Bernoulli trials, conditionally on some block-specific probabilities, collected in a $H \times H$ symmetric matrix Ξ. The generic element $\xi_{hh'}$ of Ξ represents the probability of occurrence of an edge between a node in community h and a node in the community h'. More precisely, encoding the partition into communities through a vector $\mathbf{z} = (z_1, \ldots, z_V)$, where $z_u \in [H]$, the labels follow the order of appearance, and $z_u = z_v$ if and only if the nodes u and v belong to the same community, then the classical SBM is defined by

$$(y_{uv})_{u<v} \mid z_u = h, z_v = h', \Xi \overset{ind}{\sim} \text{Bernoulli}(\xi_{hh'}) \tag{1}$$

Note that, since $\mathbf{Y}$ is symmetric and with fixed null diagonal, we just model its upper triangular elements.

Typically, neither the block probabilities in Ξ nor the allocations to communities, encoded in $\mathbf{z}$, are observable. A standard Bayesian approach consists of employing independent beta priors for the block-probabilities in Ξ, conditionally on the allocations, such as

$$(\xi_{hh'})_{h \leq h'} \mid \mathbf{z} \overset{iid}{\sim} \text{beta}(a, b) \tag{2}$$

which entail conjugacy with the Bernoulli likelihood. While a number of different prior can be employed for the partition $\mathbf{z}$: Dirichlet-multinomial, Dirichlet process, mixture of finite partitions, or general Gibbs-type, as in [11, 13, 16, 19], respectively. Nonetheless, direct generalizations of these priors that can accommodate temporal dynamics are not straightforward.

In the last decade, a number of extensions of the original stochastic block model has been proposed. See [15] for a comprehensive review. In [4], a recent work related to this one, the authors introduce a novel class of partially exchangeable stochastic block models (pEx-SBM) specified in terms of a hierarchical random partition prior for the allocation of nodes. In particular, for a multilayer node-colored network, defined as in [14], the prior on the random allocation $\mathbf{z}$ is induced by the ties in the realization of a latent random attribute array, whose rows follow the division in layers of the network, and that is directed by a vector of hierarchical normalized completely random measures (H-NRMIs). In other words, for what concerns such latent attribute, and, as a consequence, the random partition induced by it, nodes are modeled as partially exchangeable with respect to the division in layers. See [4] for the implications of such modeling assumption in multilayer networks. Even if the type of networks and the data are rather different from the one considered in our work, since in [4] the networks have, as said, a node-colored structure, the prior distribution and the learning mechanism are strictly connected. In Sect. 3, we describe how one can leverage the class of priors studied in [4] to model the conditional distribution of present communities given the past community structure.

2.2 Conditional Partial Exchangeability

Partial exchangeability, proposed by de Finetti in [6], is a form of distributional invariance that generalizes exchangeability to match situations in which observations are naturally divided in groups. A classical practical example is given by data related to patients treated in different hospitals. The goal of a partially exchangeable model is to induce homogeneity within groups, and allow for heterogeneity across groups, while still borrowing information. The extreme cases are represented by (full) exchangeability, where no heterogeneity is induced, and group independence, where no across-groups borrowing is allowed.

Let $(X_n)_{n\geq 1}$ be a sequence of random variables and $\ell = (\ell_n)_{n\geq 1}$, with $\ell_n \in [d]$, an allocation sequence that encodes their division in d groups. Then $(X_n)_{n\geq 1}$ is partially exchangeable with respect to ℓ if, for any $n \in \mathbb{N}$,

$$(X_{i_1}, \dots, X_{i_n}) \stackrel{d}{=} (X_{\sigma(i_1)}, \dots, X_{\sigma(i_n)}) \tag{3}$$

for any choice of indices $(i_1, \dots, i_n)$ and any $\sigma \in \mathscr{P}(n; \ell)$, the set of permutations of n elements that *preserves* ℓ. This means that $\ell_{\sigma(i_k)} = \ell_{i_k}$ for $k \in [n]$. In other words, the distribution of any finite vector of elements of $(X_n)_{n\geq 1}$ is invariant with respect to permutations within groups, but not necessarily across groups. Notice that if $d = 1$, partial exchangeability coincides with (full) exchangeability. It can be proved that (3) is equivalent to asking, for any $n \in \mathbb{N}$,

$$(X_1, \dots, X_n) \stackrel{d}{=} (X_{\sigma(1)}, \dots, X_{\sigma(n)}) \tag{4}$$

for any $\sigma \in \mathscr{P}(n; \ell)$.

Conditional partial exchangeability (CPE) was recently introduced in [9] as a modeling principle for dependent Bayesian clustering based on mixtures. The referenced data analysis problem for CPE involves detecting multiple latent clustering structures. Each of these structures partition the same set of statistical units, reflecting information related to a different feature. Let $X_i^{(1)}$ and $X_i^{(2)}$ be two observed features of the i-th item in a sample to be clustered according to them. We denote with $\mathbf{z}^{(1)}$ and $\mathbf{z}^{(2)}$ the two allocation vectors encoding the random clustering structures underlying

$X^{(1)}$ and $X^{(2)}$, respectively. The CPE modeling principle requires that, for any fixed number n of items in the sample, and any measurable set A

$$\mathbb{P}\left[(X_1^{(2)}, \ldots, X_n^{(2)}) \in A \,\middle|\, \mathbf{z}^{(1)}\right] = \mathbb{P}\left[(X_{\sigma(1)}^{(2)}, \ldots, X_{\sigma(n)}^{(2)}) \in A \,\middle|\, \mathbf{z}^{(1)}\right]$$

for any $\sigma \in \mathscr{P}(n;\ \mathbf{z}^{(1)})$. In other words, $X^{(2)}$ is *conditionally* partially exchangeable with respect to the *random* partition underlying $X^{(1)}$. As a consequence, we say that the random partition $\mathbf{z}^{(2)}$ is conditionally partially exchangeable with respect to the random partition $\mathbf{z}^{(1)}$. In [9], as said, this modeling principle is presented through mixture models, meaning that the features $X^{(1)}$ and $X^{(2)}$ arise from random mixture densities, with almost surely discrete mixing measures, which induce the partitions $\mathbf{z}^{(1)}$ and $\mathbf{z}^{(2)}$.

In this paper, we adapt CPE to stochastic block models, as described in Sect. 3. When applied to node clustering in a network, CPE will imply a conditional invariance of the predictive distribution of the network at a future time point only with respect to permutations of nodes that preserve the present-time community. See [9] for how CPE induces a dependence between the random clustering structures that takes into account the specific allocation of each item and, thus, the appropriate learning mechanism for multi-view and panel data.

3 A CPE Model for Clustering in Dynamic Networks

For the sake of simplicity, we present here the model for only two time points. A possible straightforward extension to any number of time points can be developed through a Markovian structure. Let us consider a set of nodes denoted as $1, \ldots, V$ that exhibit varying connections at two distinct time points, $t = 1$ and $t = 2$. At each time point, the network's connections are described by the binary and symmetric adjacency matrix $\mathbf{Y}^{(t)}$, having dimensions $V \times V$. The entry $y_{uv}^{(t)}$ is assigned a value of 1 if an edge exists between node v and node u at time t, for $1 \leq u < v \leq V$ and $t = 1, 2$. Our modeling approach employs a stochastic block likelihood for each adjacency matrix, contingent upon time-specific latent partitions $\mathbf{z}^{(t)}$, $t = 1, 2$. The joint distribution of the adjacency matrices and community structures can be decomposed through a telescopic clustering scheme, as introduced in [9]:

$$p\left(\mathbf{Y}^{(1)}, \mathbf{Y}^{(2)}, \mathbf{z}^{(1)}, \mathbf{z}^{(2)}\right) = \\ = p\left(\mathbf{Y}^{(1)} \mid \mathbf{z}^{(1)}\right) \cdot p\left(\mathbf{Y}^{(2)} \mid \mathbf{z}^{(2)}\right) \cdot p\left(\mathbf{z}^{(2)} \mid \mathbf{z}^{(1)}\right) \cdot p\left(\mathbf{z}^{(1)}\right) \tag{5}$$

The density $p\left(\mathbf{Y}^{(t)} \mid \mathbf{z}^{(t)}\right)$ for $t = 1, 2$ is defined as

$$p(\mathbf{Y}^{(t)} \mid \mathbf{z}^{(t)}) = \prod_{h=1}^{H^{(t)}} \prod_{h'=1}^{h} \frac{\mathrm{B}(a^{(t)} + m_{hh'}^{(t)}, b^{(t)} + \overline{m}_{hh'}^{(t)})}{\mathrm{B}(a^{(t)}, b^{(t)})} \tag{6}$$

where $m_{hh}^{(t)}$ and $\overline{m}_{hh'}^{(t)}$ are the total number of edges and non-edges among the nodes in groups h and h' of partition $\mathbf{z}^{(t)}$, respectively, $a^{(t)}, b^{(t)} > 0$ for $t = 1, 2$ and $\mathrm{B}(\cdot, \cdot)$ denotes the beta function.

As common practice in Bayesian nonparametric clustering, we attach an entry of auxiliary random vectors $\mathbf{x}^{(t)}$, for $t = 1, 2$, to each item to be clustered. Choosing such vectors to be directed by discrete random probability measures, we can induce the random partitions $\mathbf{z}^{(t)}$, for $t = 1, 2$, through the ties observed in $\mathbf{x}^{(t)}$.

In our specific case, we let $\mathbf{x}^{(1)} = (x_1^{(1)}, \ldots, x_V^{(1)})$ be such that

$$x_1^{(1)}, \ldots, x_V^{(1)} \mid \tilde{P}^{(1)} \overset{\text{iid}}{\sim} \tilde{P}^{(1)} \tag{7}$$

for some almost surely discrete random probability measure $\tilde{P}^{(1)}$, and $\mathbf{z}^{(1)}$ be the allocation vector representing the random partition induced by the ties in $\mathbf{x}^{(1)}$, and whose labels reflect the order of appearance of the unique values of $\mathbf{x}^{(1)}$. Let now $\mathbf{x}^{(2)} = (x_1^{(2)}, \ldots, x_V^{(2)})$ such that

$$x_v^{(2)} \mid (\tilde{P}_1^{(2)}, \ldots, \tilde{P}_{H^{(1)}}^{(2)}), z_v^{(1)} \overset{\text{ind}}{\sim} \tilde{P}_{z_v^{(1)}}^{(2)} \tag{8}$$

with conditional independence across $v = 1, \ldots, V$, where $(\tilde{P}_1^{(2)}, \ldots, \tilde{P}_{H^{(1)}}^{(2)})$ is a vector of discrete random probability measures and $H^{(1)}$ is the (random) number of clusters in $\mathbf{z}^{(1)}$. We denote with $\mathbf{z}^{(2)}$ the allocation vector representing the partition induced by the ties in $\mathbf{x}^{(2)}$. In this way, by de Finetti's representation theorem for partially exchangeable arrays, the dependence structure of the partitions $(\mathbf{z}^{(1)}, \mathbf{z}^{(2)})$ is indeed characterized by a CPE scheme: $\mathbf{x}^{(2)}$ is conditionally partially exchangeable with respect to the partition induced by the ties in $\mathbf{x}^{(1)}$.

When $\tilde{P}^{(1)}$ is a NRMI with Lévy intensity $\nu^{(1)}$ (see [21]) and $(\tilde{P}_1^{(2)}, \ldots, \tilde{P}_{H^{(1)}}^{(2)})$ is, conditionally on $H^{(1)}$, a vector of H-NRMI with Lévy intensities $\nu_0^{(2)}, \nu^{(2)}$ (see [2]), then we denote the distribution of $(\mathbf{z}^{(1)}, \mathbf{z}^{(2)})$ with

$$\mathbf{z}^{(2)} \mid \mathbf{z}^{(1)} \sim \mathrm{pExP}(\mathbf{z}^{(1)}; \nu_0^{(2)}, \nu^{(2)}) \tag{9}$$

$$\mathbf{z}^{(1)} \sim \mathrm{ExP}(V; \nu^{(1)}) \tag{10}$$

The notation in (9) refers to the fact that the law of the random partition induced by the vector of H-NRMIs is, conditionally on $\mathbf{z}^{(1)}$, a *partially exchangeable random partition*. The prior in (10), for analogy, indicates the law of the *exchangeable random partition* of V items induced by a NRMI. As discussed in [2] and leveraged in [4], the

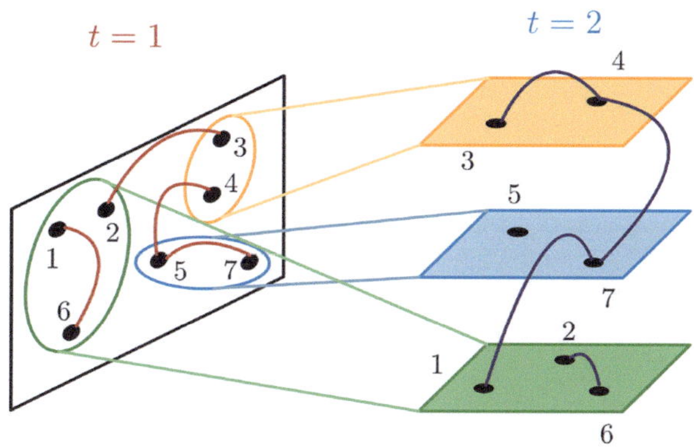

Fig. 1 A conditionally partially exchangeable network: the latent clustering at time 1 becomes the division in layers at time 2

conditional distribution of $\mathbf{z}^{(2)} \mid \mathbf{z}^{(1)}$ and the distribution of $\mathbf{z}^{(1)}$ are characterized by a partially exchangeable partition probability function (pEPPF) and an exchangeable partition probability function (EPPF), respectively.

To sum up, the likelihood in (1), completed with the priors in (9) and (10), together with the dependence structure given in (5), define our model. In the following we underline some aspects of the model formulation and assess the desired invariance property.

Remark 1 (*Conditional pEx-SBM*) The conditional distribution $p(\mathbf{Y}^{(2)}, \mathbf{z}^{(2)} \mid \mathbf{z}^{(1)})$ is described by the partially exchangeable stochastic block models (pEx-SBMs) introduced in [4]. In words, the partition at time 1 becomes at time 2 the division in layers of a (static) multilayer stochastic block model, whose prior on the latent partition is partially exchangeable with respect to the division in layers, as showed in Fig. 1. Being driven by a H-NRMI vector, the latent partition encoded in $\mathbf{z}^{(2)}$ features both within- and across-layer clusters. Hence $\mathbf{z}^{(2)}$ is informed by $\mathbf{z}^{(1)}$, but not constrained in the spaces of subpartitions or superpartions of $\mathbf{z}^{(1)}$.

Remark 2 (*Conditional independence*) In (5), the joint distribution $p(\mathbf{Y}^{(1)}, \mathbf{Y}^{(2)} \mid \mathbf{z}^{(1)}, \mathbf{z}^{(2)})$ factorizes in $p(\mathbf{Y}^{(1)} \mid \mathbf{z}^{(1)}) \cdot p(\mathbf{Y}^{(2)} \mid \mathbf{z}^{(2)})$. This implies a hidden Markov model representation where conditional independence between time points arises when we condition on the latent partitions $\mathbf{z}^{(1)}$ and $\mathbf{z}^{(2)}$. In particular we have that $\mathbf{Y}^{(2)} \mid \mathbf{z}^{(2)} \perp\!\!\!\perp \mathbf{z}^{(1)} \mid \mathbf{z}^{(2)}$, which is a feature of the pEx-SBMs: given the latent partition, there is no modeling difference between an across-layer edge and a within-layer one. In fact, at time 2, the information about the division in layers, i.e. the past partition, is solely in the prior on the present partition. This clean separation between structural information and observed connections allows for a certain degree of prior elicitation: as showed in [4], it is possible to choose the hyperparameters of the selected H-NRMI to obtain a less or greater adherence of the latent partition to the division in layer. Translated in dynamic terms, this results allow for tuning the similarity of the present partition to the past one.

Remark 3 (*Invariance*) It is an immediate consequence of the representation in Theorem 1 of [9] that the law of $(\mathbf{x}^{(1)}, \mathbf{x}^{(2)})$ possesses the following invariance

$$(x_1^{(1)}, \ldots, x_V^{(1)}, x_1^{(2)}, \ldots, x_V^{(2)}) \stackrel{d}{=} (x_{\sigma(1)}^{(1)}, \ldots, x_{\sigma(V)}^{(1)}, x_{\sigma(1)}^{(2)}, \ldots, x_{\sigma(V)}^{(2)}) \tag{11}$$

for any permutation σ of V elements. In other words, if we consider the matrix $\begin{pmatrix}\mathbf{x}^{(1)}\\ \mathbf{x}^{(2)}\end{pmatrix}$, its joint distribution is invariant with respect to column swapping.

From Theorem 2 in [9], instead, we obtain that, in general,

$$(x_1^{(1)}, \ldots, x_V^{(1)}, x_1^{(2)}, \ldots, x_V^{(2)}) \stackrel{d}{\neq} (x_1^{(2)}, \ldots, x_V^{(2)}, x_1^{(1)}, \ldots, x_V^{(1)}) \tag{12}$$

and

$$(x_1^{(1)}, \ldots, x_V^{(1)}, x_1^{(2)}, \ldots, x_V^{(2)}) \stackrel{d}{\neq} (x_{\sigma(1)}^{(1)}, \ldots, x_{\sigma(V)}^{(1)}, x_{\sigma'(1)}^{(2)}, \ldots, x_{\sigma'(V)}^{(2)}) \tag{13}$$

for $\sigma \neq \sigma'$ permutations of V elements. In words, (12) implies that the rows of the matrix $\begin{pmatrix}\mathbf{x}^{(1)}\\ \mathbf{x}^{(2)}\end{pmatrix}$ are not exchangeable, i.e. we cannot swap nodes in different times. On the other hand, (13) excludes partial exchangeability across times, that is we cannot swap nodes in the same time point, if the same permutation is not coherently applied to every time point. Therefore, CPE induces the appropriate invariance property for dynamic clustering: the joint distribution and the inference remain unchanged under node labels swapping if and only if the swap is not across times and it is coherently applied to all the instances of the same node in different times. This invariance structure ensures preservation of the node identity and, as a consequence, entails that all information in the data is leveraged to conduct inference, leading to a more accurate representation of node correspondence across time. This is particularly important in scenarios where the identity of the nodes has a significant impact on the analysis.

Remark 4 (*Projectivity*) Again from the representation in Theorem 1 of [9], we deduce that the proposed model is Kolmogorov consistent in V. This implies that if m new nodes join the dynamic network, their connections can be modeled conditionally on the connections among the old V nodes, and hence coherently predicted.

Remark 5 (*Temporal dependence*) The likelihood in (6) is a beta-Bernoulli probability function, obtained integrating out the cluster-specific connection probabilities $\Xi^{(t)} = (\xi_{hk}^{(t)})_{h,k=1,\ldots,H^{(t)}}$ for $t = 1, 2$. Such probabilities, are, at each time, given the latent partition, independent draws form a beta distribution. One may think that it would be reasonable to inject some across-time dependence in the connection probabilities to reinforce the temporal structure. However, this is a non-trivial modeling task. On one hand, the cluster labels are non-identifiable objects that cannot be interpreted, hence one cannot link present probabilities to specific (couple of) past clusters, but only to the nodes that are allocated in those past clusters. On the other hand, creating dependence between the present clustering probability of a certain node and past connection probabilities of other nodes evidently breaks Kolmogorov consistency. See [8] for a change of viewpoint that allows for overcoming this issue.

4 Posterior Inference

The distribution of interest is $p(\mathbf{z}^{(1)}, \mathbf{z}^{(2)} \mid \mathbf{Y}^{(1)}, \mathbf{Y}^{(2)})$. Using the same notation as in [4], we target the augmented posterior $p(\mathbf{z}^{(1)}, \mathbf{z}^{(2)}, \mathbf{w} \mid \mathbf{Y}^{(1)}, \mathbf{Y}^{(2)})$, where $\mathbf{w}$ denotes the allocation vector representing the nested partition induced by the hierarchical pExP prior. Phrasing it in the well-known Chinese restaurant franchise metaphor, introduced in [24], at time 2 the partition in $\mathbf{z}^{(1)}$ gives the division in restaurants, the partition in $\mathbf{z}^{(2)}$ is the across-restaurants division in eaten dishes, while the nested partition in $\mathbf{w}$ represents the division in tables. Notice that $\mathbf{w}$ is nested in both $\mathbf{z}^{(1)}$ and $\mathbf{z}^{(2)}$. Taking a conditional pEx-SBM viewpoint, exemplified in Fig. 1, we refer to the elements of the nested partition in $\mathbf{w}$ as *subclusters*, to the elements of the partition in $\mathbf{z}^{(1)}$ as *layers*, and to the elements of the partition in $\mathbf{z}^{(2)}$ as *clusters*. Let us denote with ℓ_{kh} the number of subclusters in the intersection of layer k of $\mathbf{z}^{(1)}$ and cluster h of $\mathbf{z}^{(2)}$, with $\ell_{k\cdot} := \sum_{h=1}^{H^{(2)}} \ell_{kh}$ the number of subclusters in layer k of $\mathbf{z}^{(1)}$, with the vector $\mathbf{q}_{kh} = (q_{kh1}, \dots, q_{kh\ell_{kh}})$ the number of nodes in layer k and cluster h for each subcluster, with $(\mathbf{q}_{k1}, \dots, \mathbf{q}_{kH^{(2)}})$ the simple juxtaposition of the vectors $\mathbf{q}_{kh}$ for $h = 1, \dots, H^{(2)}$, and with $n_k^{(1)}$ the number of nodes in layer k. The apex $^{-v}$ indicates that a quantity has been evaluated without considering node v.

We devise a node-wise blocked Gibbs sampler, that is for any $v \in [V]$, we sample from

$$\mathbb{P}(z_v^{(1)} = k, z_v^{(2)} = h, w_v = \tau \mid \mathbf{z}^{(2)-v}, \mathbf{w}^{-v}, \mathbf{z}^{(1)-v}, \mathbf{Y}^{(1)}, \mathbf{Y}^{(2)}) \tag{14}$$

for any $k \in [H^{(1)-v} + 1]$, $h \in [H^{(2)-v} + 1]$, and $\tau \in [\ell_{k\cdot}^{-v} + 1]$. We have that (14) is equal to

$$\frac{p(\mathbf{Y}^{(1)}, \mathbf{Y}^{(2)} \mid z_v^{(1)} = k, z_v^{(2)} = h, w_v = \tau, \mathbf{z}^{(1)-v}, \mathbf{z}^{(2)-v}, \mathbf{w}^{-v})}{p(\mathbf{Y}^{(1)}, \mathbf{Y}^{(2)} \mid \mathbf{z}^{(1)-v}, \mathbf{z}^{(2)-v}, \mathbf{w}^{-v})} \times$$
$$\times\, \mathbb{P}\left(z_v^{(1)} = k, z_v^{(2)} = h, w_v = \tau \,\middle|\, \mathbf{z}^{(2)-v}, \mathbf{w}^{-v}, \mathbf{z}^{(1)-v}\right) \propto$$
$$\propto \prod_{t=1}^{2} \prod_{h'=1}^{H^{(t)-v}} \frac{\mathrm{B}(a^{(t)} + m_{hh'}^{(t)-v} + r_{vh'}^{(t)}, b^{(t)} + \overline{m}_{hh'}^{(t)-v} + \overline{r}_{vh'}^{(t)})}{\mathrm{B}(a^{(t)} + m_{hh'}^{(t)-v}, b^{(t)} + \overline{m}_{hh'}^{(t)-v})} \times$$
$$\times\, \mathbb{P}\left(z_v^{(1)} = k, z_v^{(2)} = h, w_v = \tau \,\middle|\, \mathbf{z}^{(2)-v}, \mathbf{w}^{-v}, \mathbf{z}^{(1)-v}\right)$$

where $r_{vh'}^{(t)}$ and $\overline{r}_{vh'}^{(t)}$ denote the number of edges and non-edges between node v and nodes in layer h', if $t = 1$, or in cluster h', if $t = 2$. Hence the likelihood part can be easily calculated. We treat the prior part leveraging the conditional representation and the general urn schemes induced by H-NRMIs and NRMIs. We have

$$\mathbb{P}\left(z_v^{(1)} = k,\ z_v^{(2)} = h,\ w_v = \tau \left| \mathbf{z}^{(2)-v},\ \mathbf{w}^{-v},\ \mathbf{z}^{(1)-v}\right.\right) =$$
$$= \frac{p(z_v^{(1)} = k,\ z_v^{(2)} = h,\ w_v = \tau,\ \mathbf{z}^{(2)-v},\ \mathbf{w}^{-v},\ \mathbf{z}^{(1)-v})}{p(\mathbf{z}^{(2)-v},\ \mathbf{w}^{-v},\ \mathbf{z}^{(1)-v})} =$$
$$= \frac{p(z_v^{(2)} = h,\ w_v = \tau,\ \mathbf{z}^{(2)-v},\ \mathbf{w}^{-v} \mid z_v^{(1)} = k,\ \mathbf{z}^{(1)-v})}{p(\mathbf{z}^{(2)-v},\ \mathbf{w}^{-v} \mid \mathbf{z}^{(1)-v})} \mathbb{P}(z_v^{(1)} = k \mid \mathbf{z}^{(1)-v}) \quad (15)$$

where

$$\mathbb{P}(z_v^{(1)} = k \mid \mathbf{z}^{(1)-v}) = \frac{{}_{(1)}\Phi^{(V)}_{H_k^{(1)-v}}\left(n_1^{(1)-v}, \ldots, n_k^{(1)-v} + 1, \ldots, n_{H_k^{(1)-v}}^{(1)-v}\right)}{{}_{(1)}\Phi^{(V-1)}_{H^{(1)-v}}\left(n_1^{(1)-v}, \ldots, n_k^{(1)-v}, \ldots, n_{H^{(1)-v}}^{(1)-v}\right)}$$

with ${}_{(1)}\Phi^{(\cdot)}_{\cdot}$ being the EPPF relative to the Lévy intensity $\nu^{(1)}$, with the notation adopted in [2, 4], and $H_k^{(1)-v} = \max\{H^{(1)-v}, k\}$. Following the reasoning in the proof of Proposition 3.1 in [4], we can retrieve numerator and denominator of (15). For any $k \in [H^{(1)-v}]$, we need to evaluate the numerator for every couple of values $(h, \tau) \in [H^{(2)-v} + 1] \times [\ell_{k\cdot}^{-v} + 1]$, but just few of these entries are non-zero. In fact, for any subcluster $\tau \in [\ell_{k\cdot}^{-v}]$, the associated cluster is fixed. We define $\mathbb{T}_{kh}^{-v}$ as in [4]: in words, it is the set of the indices, among those counting the subclusters in layer k, relative to subclusters in cluster h, for any $h \in \mathbf{H}_k^{(2)-v}$, which is the set of unique cluster labels in layer k, as in [4]. Let $\boldsymbol{\tau}_{kh}^{-v}$ be the vector obtained ordering $\mathbb{T}_{kh}^{-v}$. Then for any $h \in [H^{(2)-v}]$ and $s \in [\ell_{kh}]$ we have

$$p(z_v^{(2)} = h,\ w_v = \tau_{khs}^{-v},\ \mathbf{z}^{(2)-v},\ \mathbf{w}^{-v} \mid z_v^{(1)} = k,\ \mathbf{z}^{(1)-v}) =$$
$$= {}_{(2)}\Phi^{(|\boldsymbol{\ell}^{-v}|)}_{H^{(2)-v},0}(\ell_{\cdot 1}^{-v}, \ldots, \ell_{\cdot H^{(2)-v}}^{-v}) \prod_{\substack{k'=1\\k'\neq k}}^{H^{(1)-v}} {}_{(2)}\Phi^{(n_{k'}^{-v})}_{\ell_{k'\cdot}^{-v},k'}(\mathbf{q}_{k'1}^{-v}, \ldots, \mathbf{q}_{k'H^{(2)-v}}^{-v}) \times$$
$$\times\ {}_{(2)}\Phi^{(n_k^{-v}+1)}_{\ell_{k\cdot}^{-v},k}(\mathbf{q}_{k1}^{-v}, \ldots, \mathbf{q}_{kh}^{-v} + \boldsymbol{e}_s, \ldots, \mathbf{q}_{kH^{(2)-v}}^{-v})$$

where $\boldsymbol{e}_s = (0, \ldots 0, 1, 0 \ldots, 0)$ and the 1 is in the s-th entry. Instead for any $h \in [H^{(2)-v} + 1]$ and $\tau = \ell_{k\cdot}^{-v} + 1$ we have

$$p(z_v^{(2)} = h,\ w_v = \ell_{k\cdot}^{-v} + 1,\ \mathbf{z}^{(2)-v},\ \mathbf{w}^{-v} \mid z_v^{(1)} = k,\ \mathbf{z}^{(1)-v}) =$$
$$= {}_{(2)}\Phi^{(|\boldsymbol{\ell}^{-v}|+1)}_{H_h^{(2)-v},0}(\ell_{\cdot 1}^{-v}, \ldots, \ell_{\cdot h}^{-v} + 1, \ldots, \ell_{\cdot H_v^{(2)-v}}^{-v}) \times$$
$$\times \prod_{\substack{k'=1\\k'\neq k}}^{H^{(1)-v}} {}_{(2)}\Phi^{(n_{k'}^{-v})}_{\ell_{k'\cdot}^{-v},k'}(\mathbf{q}_{k'1}^{-v}, \ldots, \mathbf{q}_{k'H^{(2)-v}}^{-v})\ {}_{(2)}\Phi^{(n_k^{-v}+1)}_{\ell_{k\cdot}^{-v}+1,k}(\mathbf{q}_{k1}^{-v}, \ldots, (\mathbf{q}_{kh}^{-v}, 1), \ldots, \mathbf{q}_{kH_h^{(2)-v}}^{-v})$$

${}_{(2)}\Phi^{(\cdot)}_{\cdot,0}$ and ${}_{(2)}\Phi^{(\cdot)}_{\cdot,k}$ are the EPPFs relative to Lévy intensities $\nu_0^{(2)}$ and $\nu^{(2)}$ respectively, and $H_h^{(2)-\nu} = \max\{H^{(2)-\nu}, h\}$. Finally for $k = H^{(1)-\nu} + 1$, the ν-th node is assigned to an unobserved layer. Hence just $\tau = 1$ has non-zero probability, and we have

$$p(z_\nu^{(2)} = h,\ w_\nu = 1,\ \mathbf{z}^{(2)-\nu},\ \mathbf{w}^{-\nu} \mid z_\nu^{(1)} = H^{(1)-\nu} + 1,\ \mathbf{z}^{(1)-\nu}) =$$
$$= {}_{(2)}\Phi^{(|\boldsymbol{\ell}^{-\nu}|+1)}_{H_h^{(2)-\nu},0}(\ell_{\cdot 1}^{-\nu}, \ldots, \ell_{\cdot h}^{-\nu} + 1, \ldots, \ell^{-\nu}_{\cdot H_\nu^{(2)-\nu}}) \times$$
$$\times \prod_{k'=1}^{H^{(1)-\nu}} {}_{(2)}\Phi^{(n_{k'}^{-\nu})}_{\ell_{k'\cdot}^{-\nu},\,k'}(\mathbf{q}_{k'1}^{-\nu}, \ldots, \mathbf{q}^{-\nu}_{k' H^{(2)-\nu}})\ {}_{(2)}\Phi^{(1)}_{1, H^{(1)-\nu}+1}(1)$$

for any $h \in [H^{(2)-\nu} + 1]$. The denominator is

$$p(\mathbf{z}^{(2)-\nu},\ \mathbf{w}^{-\nu} \mid \mathbf{z}^{(1)-\nu}) = {}_{(2)}\Phi^{(|\boldsymbol{\ell}^{-\nu}|)}_{H^{(2)-\nu},0}(\ell_{\cdot 1}^{-\nu}, \ldots, \ell^{-\nu}_{\cdot H^{(2)-\nu}}) \times$$
$$\times\ {}_{(2)}\Phi^{(n_k^{-\nu})}_{\ell_{k\cdot}^{-\nu},\,k}(\mathbf{q}_{k1}^{-\nu}, \ldots, \mathbf{q}^{-\nu}_{k H^{(2)-\nu}}) \prod_{\substack{k'=1\\ k'\neq k}}^{H^{(1)-\nu}} {}_{(2)}\Phi^{(n_{k'}^{-\nu})}_{\ell_{k'\cdot}^{-\nu},\,k'}(\mathbf{q}_{k'1}^{-\nu}, \ldots, \mathbf{q}^{-\nu}_{k' H^{(2)-\nu}})$$

Therefore we can state the following result.

Proposition 1 *Let* $(\mathbf{z}^{(1)}, \mathbf{z}^{(2)})$ *be a pair of random allocation vectors distributed as in* (9) *and* (10). *Then for any node* ν *we have the following:*
(1) If $k \in [H^{(1)-\nu}]$, $h \in \mathbf{H}_k^{(2)-\nu}$, *and* $s \in [\ell_{kh}^{-\nu}]$ *:*

$$\mathbb{P}\left(z_\nu^{(1)} = k,\ z_\nu^{(2)} = h,\ w_\nu = \tau_{khs}^{-\nu} \,\middle|\, \mathbf{z}^{(2)-\nu},\ \mathbf{w}^{-\nu},\ \mathbf{z}^{(1)-\nu}\right) =$$
$$= \frac{{}_{(2)}\Phi^{(n_k^{-\nu}+1)}_{\ell_{k\cdot}^{-\nu},k}(\mathbf{q}_{k1}^{-\nu}, \ldots, \mathbf{q}_{kh}^{-\nu} + \boldsymbol{e}_s, \ldots, \mathbf{q}^{-\nu}_{k H^{(2)-\nu}})}{{}_{(2)}\Phi^{(n_k^{-\nu})}_{\ell_{k\cdot}^{-\nu},\,k}(\mathbf{q}_{k1}^{-\nu}, \ldots, \mathbf{q}^{-\nu}_{k H^{-\nu}})} \times$$
$$\times \frac{{}_{(1)}\Phi^{(V)}_{H_k^{(1)-\nu}}\left(n_1^{(1)-\nu}, \ldots, n_k^{(1)-\nu} + 1, \ldots, n^{(1)-\nu}_{H_k^{(1)-\nu}}\right)}{{}_{(1)}\Phi^{(V-1)}_{H^{(1)-\nu}}\left(n_1^{(1)-\nu}, \ldots, n_k^{(1)-\nu}, \ldots, n^{(1)-\nu}_{H^{(1)-\nu}}\right)} \tag{16}$$

where $\tau_{khs}^{-\nu}$ *is the* s*-th entry of the vector* $\boldsymbol{\tau}_{kh}^{-\nu}$ *obtained ordering the set* $\mathbb{T}_{kh}^{-\nu}$, *and* $H_k^{(1)-\nu} = H^{(1)-\nu} \vee k$.

(2) If $k \in [H^{(1)-v}]$, $\tau = \ell_{k\cdot}^{-v} + 1$ *and* $h \in [H^{(2)-v} + 1]$*:*

$$\mathbb{P}(z_v^{(1)} = k,\ z_v^{(2)} = h,\ w_v = \ell_{k\cdot}^{-v} + 1 \mid \mathbf{z}^{(1)-v},\ \mathbf{z}^{(2)-v},\ \mathbf{w}^{-v}) =$$

$$= \frac{{}_{(2)}\Phi^{(|\boldsymbol{\ell}^{-v}|+1)}_{H_h^{(2)-v},0}(\ell_{\cdot 1}^{-v}, \ldots, \ell_{\cdot h}^{-v} + 1, \ldots, \ell^{-v}_{\cdot H_v^{(2)-v}})}{{}_{(2)}\Phi^{(|\boldsymbol{\ell}^{-v}|)}_{H^{(2)-v},0}(\ell_{\cdot 1}^{-v}, \ldots, \ell^{-v}_{\cdot H^{(2)-v}})} \times$$
$$\times \frac{{}_{(2)}\Phi^{(n_k^{-v}+1)}_{\ell_{k\cdot}^{-v}+1,k}(\mathbf{q}_{k1}^{-v}, \ldots, (\mathbf{q}_{kh}^{-v}, 1), \ldots, \mathbf{q}^{-v}_{kH_h^{(2)-v}})}{{}_{(2)}\Phi^{(n_k^{-v})}_{\ell_{k\cdot}^{-v},k}(\mathbf{q}_{k1}^{-v}, \ldots, \mathbf{q}_{kH^{-v}}^{-v})} \times$$
$$\times \frac{{}_{(1)}\Phi^{(V)}_{H_k^{(1)-v}}\left(n_1^{(1)-v}, \ldots, n_k^{(1)-v} + 1, \ldots, n^{(1)-v}_{H_k^{(1)-v}}\right)}{{}_{(1)}\Phi^{(V-1)}_{H^{(1)-v}}\left(n_1^{(1)-v}, \ldots, n_k^{(1)-v}, \ldots, n^{(1)-v}_{H^{(1)-v}}\right)} \tag{17}$$

where $H_h^{(2)-v} = H^{(2)-v} \vee h$.
(3) If $k = H^{(1)-v} + 1$ *and* $h \in [H^{(2)-v} + 1]$*:*

$$\mathbb{P}(z_v^{(1)} = H^{(1)-v} + 1,\ z_v^{(2)} = h,\ w_v = 1 \mid \mathbf{z}^{(1)-v},\ \mathbf{z}^{(2)-v},\ \mathbf{w}^{-v}) =$$

$$= \frac{{}_{(2)}\Phi^{(|\boldsymbol{\ell}^{-v}|+1)}_{H_h^{(2)-v},0}(\ell_{\cdot 1}^{-v}, \ldots, \ell_{\cdot h}^{-v} + 1, \ldots, \ell^{-v}_{\cdot H_v^{(2)-v}})}{{}_{(2)}\Phi^{(|\boldsymbol{\ell}^{-v}|)}_{H^{(2)-v},0}(\ell_{\cdot 1}^{-v}, \ldots, \ell^{-v}_{\cdot H^{(2)-v}})} \, {}_{(2)}\Phi^{(1)}_{1,H^{(1)-v}+1}(1) \times$$
$$\times \frac{{}_{(1)}\Phi^{(V)}_{H_k^{(1)-v}}\left(n_1^{(1)-v}, \ldots, n_k^{(1)-v} + 1, \ldots, n^{(1)-v}_{H_k^{(1)-v}}\right)}{{}_{(1)}\Phi^{(V-1)}_{H^{(1)-v}}\left(n_1^{(1)-v}, \ldots, n_k^{(1)-v}, \ldots, n^{(1)-v}_{H^{(1)-v}}\right)} \tag{18}$$

(4) For any other choice of $(k,\ h) \in [H^{(1)-v} + 1] \times [H^{(2)-v} + 1]$ *and* $\tau \in [\ell_{k\cdot}^{-v} + 1]$ *the probability is* 0.

5 Conclusions and Further Work

In this paper we propose a dynamic stochastic block model that *(i)* automatically learns the number of communities at each time, *(ii)* induces a probabilistic invariance that matches the complex structure of the network, *(iii)* is projective, in the Kolmogorov sense, *(iv)* allows for exact posterior inference through a node-wise blocked Gibbs sampler. Fulfilling *(i)–(iv)*, this proposal is competitive in the vast literature of models for dynamic networks. Thanks to *(i)*, we have a flexible model that does not need *ad hoc* techniques for the selection of the number of communities. Because of *(ii)*, we can include this proposal in a collection of studies [4, 8] operating under the overarching idea that structural information encoded in networks, like edge or node characteristics, can be induced in the model through the distributional symmetries implied by the prior. As mentioned, *(iii)* implies that the model is ready-made for predicting the connectivity behavior of unobserved batches of nodes. This allows, in principle, for performance comparisons with network models and algorithmic techniques that are very different in formulation. For this reason, posterior predictive schemes represent crucial future developments. The presented algorithm needs to be specialized to particular choices of NRMIs and H-NRMIs so that its performances can be evaluated against competitors, and this is also object of further work.

References

1. Abbe, E.: Community detection and stochastic block models: recent developments. J. Mach. Learn. Res. **18**(1), 6446–6531 (2017)
2. Camerlenghi, F., Lijoi, A., Orbanz, P., Prünster, I.: Distribution theory for hierarchical processes. Ann. Stat. **47**(1), 67–92 (2019)
3. DuBois, C., Butts, C., Smyth, P.: Stochastic blockmodeling of relational event dynamics. In: Carvalho, C.M., Ravikumar, P. (eds.) Proceedings of the Sixteenth International Conference on Artificial Intelligence and Statistics, vol. 31, pp. 238–246 (2013)
4. Durante, D., Gaffi, F., Lijoi, A., Prünster, I.: Partially Exchangeable Stochastic Block Models for (Node-Colored) Multilayer Networks. J. Am. Stat. Assoc. (In press) (2025)
5. Fan, X., Cao, L., Xu, R.Y.D.: Dynamic infinite mixed-membership stochastic blockmodel. IEEE Trans. Neural Netw. Learn. Syst. **26**, 2072–2085 (2013)
6. de Finetti, B.: Sur la condition d'équivalence partielle. Actual. Sci. Ind. **739**, 5–18 (1938)
7. Fortunato, S., Hric, D.: Community detection in networks: a user guide. Phys. Rep. **659**, 1–44 (2016)
8. Franzolini, B., Gaffi, F.: An invariance-based approach to dynamic block detection in networks. In preparation (2024)
9. Franzolini, B., De Iorio, M., Eriksson, J.: Conditional partial exchangeability: a probabilistic framework for multi-view clustering (2023). arXiv:2307.01152
10. Fu, W., Song, L., Xing, E.P.: Dynamic mixed membership blockmodel for evolving networks. In: Proceedings of the 26th Annual International Conference on Machine Learning, pp. 329–336 (2009)
11. Geng, J., Bhattacharya, A., Pati, D.: Probabilistic community detection with unknown number of communities. J. Am. Stat. Assoc. **114**(526), 893–905 (2019)
12. Holland, P.W., Laskey, K.B., Leinhardt, S.: Stochastic blockmodels: first steps. Soc. Netw. **5**(2), 109–137 (1983)
13. Kemp, C., Tenenbaum, J.B., Griffiths, T.L., Yamada, T., Ueda, N.: Learning systems of concepts with an infinite relational model. In: Cohn, A. (ed.) Proceedings of the 21th National Conference on Artificial Intelligence, vol. 1, pp. 381–388 (2006)
14. Kivelä, M., Arenas, A., Barthelemy, M., Gleeson, J.P., Moreno, Y., Porter, M.A.: Multilayer networks. J. Complex Netw. **2**(3), 203–271 (2014)
15. Lee, C., Wilkinson, D.J.: A review of stochastic block models and extensions for graph clustering. Appl. Netw. Sci. **4**(1), 1–50 (2019)
16. Legramanti, S., Rigon, T., Durante, D., Dunson, D.B.: Extended stochastic block models with application to criminal networks. Ann. Appl. Stat. **16**(4), 2369–2395 (2022)
17. Ludkin, M., Eckley, I.A., Neal, P.: Dynamic stochastic block models: parameter estimation and detection of changes in community structure. Stat. Comput. **28**, 1201–1213 (2018)
18. Matias, C., Miele, V.: Statistical clustering of temporal networks through a dynamic stochastic block model. J. R. Stat. Soc. Ser. B Stat. Methodol. **79**, (2015)

19. Nowicki, K., Snijders, T.A.B.: Estimation and prediction for stochastic blockstructures. J. Am. Stat. Assoc. **96**(455), 1077–1087 (2001)
20. Peixoto, T.P.: Inferring the mesoscale structure of layered, edge-valued, and time-varying networks. Phys. Rev. E **92**(4), 042807 (2015)
21. Regazzini, E., Lijoi, A., Prünster, I.: Distributional results for means of normalized random measures with independent increments. Ann. Stat. **31**(2), 560–585 (2003)
22. Schmidt, M.N., Morup, M.: Nonparametric Bayesian modeling of complex networks: an introduction. IEEE Signal Process. Mag. **30**, 110–128 (2013)
23. Tang, X., Yang, C.C.: Detecting social media hidden communities using dynamic stochastic blockmodel with temporal Dirichlet process. ACM Trans. Intell. Syst. Technol. **5**, 36:1–36:21 (2014)
24. Teh, Y.W., Jordan, M.I., Beal, M.J., Blei, D.M.: Hierarchical Dirichlet processes. J. Am. Stat. Assoc. **101**(476), 1566–1581 (2006)
25. Xing, E.P., Fu, W., Song, L.: A state-space mixed membership blockmodel for dynamic network tomography. Ann. Appl. Stat. **4**, 535–566 (2008)
26. Xu, K.S., Hero, A.O.: Dynamic stochastic blockmodels: statistical models for time-evolving networks. In: Liu, H., Salerno, J.J., Young, M.J. (eds.) Social Computing, Behavioral Modeling, and Prediction (2013)
27. Yang, T., Chi, Y., Zhu, S., Gong, Y., Jin, R.: Detecting communities and their evolutions in dynamic social networks—a Bayesian approach. Mach. Learn. **82**, 157–189 (2011)
28. Zhang, X., Moore, C., Newman, M.E.J.: Random graph models for dynamic networks. Eur. Phys. J. B **90**, 1–14 (2016)

MIX
Papier aus verantwortungsvollen Quellen
Paper from responsible sources
FSC® C105338

If you have any concerns about our products,
you can contact us on
ProductSafety@springernature.com

In case Publisher is established outside the EU,
the EU authorized representative is:
Springer Nature Customer Service Center GmbH
Europaplatz 3, 69115 Heidelberg, Germany

Printed by Libri Plureos GmbH
in Hamburg, Germany